INTRODUCTORY LECTURES ON CONVEX OPTIMIZATION

A Basic Course

Applied Optimization

Volume 87

Series Editors:

Panos M. Pardalos
University of Florida, U.S.A.

Donald W. Hearn
University of Florida, U.S.A.

INTRODUCTORY LECTURES ON CONVEX OPTIMIZATION

A Basic Course

By
Yurii Nesterov
Center of Operations Research and Econometrics, (CORE)
Université Catholique de Louvain (UCL)
Louvain-la-Neuve, Belgium

Kluwer Academic Publishers
Boston/ Dordrecht/ London

Distributors for North, Central and South America:
Kluwer Academic Publishers
101 Philip Drive
Assinippi Park
Norwell, Massachusetts 02061 USA
Telephone (781) 871-6600
Fax (781) 871-6528
E-Mail <kluwer@wkap.com>

Distributors for all other countries:
Kluwer Academic Publishers Group
Post Office Box 322
3300 AH Dordrecht, THE NETHERLANDS
Telephone 31 78 6576 000
Fax 31 78 6576 474
E-Mail <orderdept@wkap.nl>

Electronic Services <http://www.wkap.nl>

Library of Congress Cataloging-in-Publication

Nesterov, Yurri
Introductory Lectures on Convex Optimization: A Basic Course
ISBN 1-4020-7553-7

Printed on acid-free paper.

Printed in the United Kingdom by Biddles/IBT Global

Contents

Preface

It was in the middle of the 1980s, when the seminal paper by Karmarkar opened a new epoch in nonlinear optimization. The importance of this paper, containing a new polynomial-time algorithm for linear optimization problems, was not only in its complexity bound. At that time, the most surprising feature of this algorithm was that the theoretical prediction of its high efficiency was supported by excellent computational results. This unusual fact dramatically changed the style and directions of the research in nonlinear optimization. Thereafter it became more and more common that the new methods were provided with a complexity analysis, which was considered a better justification of their efficiency than computational experiments. In a new rapidly developing field, which got the name "polynomial-time interior-point methods", such a justification was obligatory.

After almost fifteen years of intensive research, the main results of this development started to appear in monographs [12, 14, 16, 17, 18, 19]. Approximately at that time the author was asked to prepare a new course on nonlinear optimization for graduate students. The idea was to create a course which would reflect the new developments in the field. Actually, this was a major challenge. At the time only the theory of interior-point methods for linear optimization was polished enough to be explained to students. The general theory of self-concordant functions had appeared in print only once in the form of research monograph [12]. Moreover, it was clear that the new theory of interior-point methods represented only a part of a general theory of convex optimization, a rather involved field with the complexity bounds, optimal methods, etc. The majority of the latter results were published in different journals in Russian.

The book you see now is a result of an attempt to present serious things in an elementary form. As is always the case with a one-semester course, the most difficult problem is the selection of the material. For us the target notions were the complexity of the optimization problems and a provable efficiency of numerical schemes supported by complexity bounds. In view of a severe volume limitation, we had to be very pragmatic. Any concept or fact included in the book is absolutely necessary for the analysis of at least one optimization scheme. Surprisingly enough, none of the material presented requires any facts from duality theory. Thus, this topic is completely omitted. This does not mean, of course, that the author neglects this fundamental concept. However, we hope that for the first treatment of the subject such a compromise is acceptable.

The main goal of this course is the development of a correct understanding of the complexity of different optimization problems. This goal was not chosen by chance. Every year I meet Ph.D. students of different specializations who ask me for advice on reasonable numerical schemes for their optimization models. And very often they seem to have come too late. In my experience, if an optimization model is created without taking into account the abilities of numerical schemes, the chances that it will be possible to find an acceptable numerical solution are close to zero. In any field of human activity, if we create something, we know *in advance* why we are doing so and what we are going to do with the result. And only in numerical modelling is the situation still different.

This course was given during several years at Université Catholique de Louvain (Louvain-la-Neuve, Belgium). The course is self-contained. It consists of four chapters (*Nonlinear optimization*, *Smooth convex optimization*, *Nonsmooth convex optimization* and *Structural optimization* (Interior-point methods)). The chapters are essentially independent and can be used as parts of more general courses on convex analysis or optimization. In our experience each chapter can be covered in three two-hour lectures. We assume a reader to have a standard undergraduate background in analysis and linear algebra. We provide the reader with short bibliographical notes which should help in a closer examination of the subject.

YURII NESTEROV

Louvain-la-Neuve, Belgium
May, 2003.

To my wife Svetlana

Acknowledgments

This book is a reflection of the main achievements in convex optimization, the field in which the author has worked for more than twenty five years. During all these years the author has had the exceptional opportunity to communicate and collaborate with the top-level scientists in the field. I am greatly indebted to many of them.

I was very lucky to start my scientific career in Moscow at the time of decline of the Soviet Union, which managed to gather in a single city the best brains of a 300-million population. The contacts with A. Antipin, Yu. Evtushenko, E. Golshtein, A. Ioffe, V. Karmanov, L. Khachian, R. Polyak, V. Pschenichnyj, N. Shor, N. Tretiakov, F. Vasil'ev, D. Yudin, and, of course, with A. Nemirovsky and B. Polyak, were invaluable in forming the directions and priorities of my research.

I was very lucky to move to the West at a very important moment in time. For nonlinear optimization that was the era of interior-point methods. That was the time, when a new paper was announced almost every day, and a time of open contacts and interesting conferences. I am very thankful to my colleges Kurt Anstreicher, Freddy Auslender, Rony Ben-Tal, Rob Freund, Jean-Louis Goffin, Don Goldfarb, Osman Guller, Florian Jarre, Ken Kortanek, Claude Lemarechal, Olvi Mangasarian, Florian Potra, Jim Renegar, Kees Roos, Tamas Terlaky, Mike Todd, Levent Tuncel and Yinyu Ye for interesting discussions and cooperation. Special thanks to Jean-Philippe Vial, the author of the idea of writing this book.

Finally, I was very lucky to find myself at the Center of Operations Research and Econometrics (CORE) in Louvain-la-Neuve, Belgium. The excellent working conditions of this research center and the exceptional environment were very helpful during all these years. It is impossible to overestimate the importance of the spirit of research, which is created and maintained here by my colleagues Vincent Blondel, Yves Genin,

Michel Gevers, Etienne Loute, Yves Poches, Yves Smeers, Paul Van Dooren and Laurence Wolsey, both coming from CORE and CESAME, a research center of the Engineering department of Université Catholique de Louvain (UCL). The research activity of the author during many years was supported by the Belgian Program on Interuniversity Poles of Attraction initiated by the Belgian State, Prime Minister's Office and Science Policy Programming.

Introduction

Optimization problems arise naturally in different fields of applications. In many situations, at some point we get a craving to arrange things in a best possible way. This intention, converted into a mathematical form, becomes an optimization problem of a certain type. Depending on the field of interest, it could be an optimal design problem, an optimal control problem, an optimal location problem, an optimal diet problem, etc. However, the next step, finding a solution to the mathematical model, is far from trivial. At first glance, everything looks very simple: many commercial optimization packages are easily available and any user can get a "solution" to the model just by clicking on an icon on the screen of his/her personal computer. The question is, what do we actually get? How much can we trust the answer?

One of the goals of this course is to show that, despite their attraction, the proposed "solutions" of general optimization problems very often cannot satisfy the expectations of a naive user. In our opinion, the main fact, which should be known to any person dealing with optimization models, is that in general *optimization problems are unsolvable.* This statement, which is usually missing in standard optimization courses, is very important for an understanding of optimization theory and its development in the past and in the future.

In many practical applications the process of creating a model can take a lot of time and effort. Therefore, the researchers should have a clear understanding of the properties of the model they are constructing. At the stage of modelling, many different tools can be used to approximate the real situation. And it is absolutely necessary to understand the computational consequences of each decision. Very often we have to choose between a "good" model, which we cannot solve,[1] and a "bad" model, which can be solved for sure. What is better?

In fact, the computational practice provides us with a hint of an answer to the above question. Actually, the most widespread optimization models now are still *linear optimization* models. It is very unlikely that such models can describe our nonlinear world very well. Thus, the main reason for their popularity is that practitioners prefer to deal with solvable models. Of course, very often the linear approximation is poor. But usually it is possible to predict the consequences of such a choice and make a correction in interpretation of the obtained solution. It seems that for them this is better than trying to solve a model without any guarantee for success.

Another goal of this course consists in discussing numerical methods for *solvable nonlinear models*, namely *convex optimization problems*. The development of convex optimization theory in the last years has been very rapid and very exciting. Now it consists of several competing branches, each of which has some strong and some weak points. We will discuss in detail their features, taking into account the historical aspect. More precisely, we will try to understand the internal logic of the development of each branch of the field. Up to now, the main results of the development can be found only in special journals and monographs. However, in our opinion, this theory is ripe for explanation to the final users, industrial engineers, economists and students of different specializations. We hope that this book will be interesting even for the experts in optimization theory since it contains many results, which have never been published in English.

In this book we will try to convince the reader, that in order to apply the optimization formulations successfully, it is necessary to be aware of some theory, which explains what we can and what we cannot do with optimization problems. The elements of this simple theory can be found in each lecture of the course. We will try to show that convex optimization is an excellent example of a *complete* application theory, which is simple, easy to learn and which can be very useful in practical applications.

In this course we discuss the most efficient modern optimization schemes and establish for them efficiency bounds. This course is self-contained; we prove all necessary results. Nevertheless, the proofs and the reasoning should not be a problem even for graduate students.

[1] More precisely, which we *can try* to solve

The structure of the book is as follows. It consists of four relatively independent chapters. Each chapter includes three sections, each of which corresponds approximately to a two-hour lecture. Thus, the contents of the book can be directly used for a standard one-semester course.

Chapter 1 is devoted to *general optimization* problems. In Section 1.1 we introduce the terminology, the notions of oracle, black box, functional model of an optimization problem and the complexity of general iterative schemes. We prove that global optimization problems are "unsolvable" and discuss the main features of different fields of optimization theory. In Section 1.2 we discuss two main local unconstrained minimization schemes: the gradient method and the Newton method. We establish their local rates of convergence and discuss the possible difficulties (divergence, convergence to a saddle point). In Section 1.3 we compare the formal structures of the gradient and the Newton method. This analysis leads to the idea of a variable metric. We describe quasi-Newton methods and conjugate gradients schemes. We conclude this section with an analysis of sequential unconstrained minimization schemes.

In **Chapter 2** we consider *smooth convex optimization* methods. In Section 2.1 we analyze the main reason for the difficulties encountered in the previous chapter and from this analysis *derive* two good functional classes, the class of smooth convex functions and that of smooth strongly convex functions. For corresponding unconstrained minimization problems we establish the lower complexity bounds. We conclude this section with an analysis of a gradient scheme, which demonstrates that this method is not optimal. The optimal schemes for smooth convex minimization problems are discussed in Section 2.2. We start from the unconstrained minimization problem. After that we introduce convex sets and define a notion of gradient mapping for a minimization problem with simple constraints. We show that the gradient mapping can formally replace a gradient step in the optimization schemes. In Section 2.3 we discuss more complicated problems, which involve several smooth convex functions, namely, the minimax problem and the constrained minimization problem. For both problems we introduce the notion of gradient mapping and present the optimal schemes.

Chapter 3 is devoted to the theory of *nonsmooth convex optimization.* Since we do not assume that the reader has a background in convex analysis, the chapter is started by Section 3.1, which contains a compact presentation of all necessary facts. The final goal of this section is to justify the rules for computing the subgradients of a convex function. The next Section 3.2 starts from the lower complexity bounds for nonsmooth optimization problems. After that we present a general scheme for the complexity analysis of the corresponding methods. We use this scheme to establish the convergence rate of the subgradient method, the center-of-gravity method and the ellipsoid method. We also discuss some other cutting plane schemes. Section 3.3 is devoted to the minimization schemes, which employ a piece-wise linear model of a convex function. We describe Kelley's method and show that it can be extremely slow. After that we introduce the so-called level method. We justify its efficiency estimates for unconstrained and constrained minimization problems.

Chapter 4 is devoted to convex minimization problems with explicit structure. In Section 4.1 we discuss a certain contradiction in the black box concept as applied to a convex optimization model. We introduce a *barrier model* of an optimization problem, which is based on the notion of *self-concordant function*. For such functions the second-order oracle is not local and they can be easily minimized by the Newton method. We study the properties of these functions and establish the rate of convergence of the Newton method. In Section 4.2 we introduce *self-concordant barriers*, the subclass of self-concordant functions, which is suitable for sequential unconstrained minimization schemes. We study the properties of such barriers and prove the efficiency estimate of the path-following scheme. In Section 4.3 we consider several examples of optimization problems, for which we can construct a self-concordant barrier, and, consequently, these problems can be solved by a path-following scheme. We consider linear and quadratic optimization problems, problems of semidefinite optimization, separable optimization and geometrical optimization, problems with extremal ellipsoids, and problems of approximation in l_p-norms. We conclude this chapter and the whole course by a comparison of an interior-point scheme with a nonsmooth optimization method as applied to a particular problem instance.

Chapter 1

NONLINEAR OPTIMIZATION

1.1 World of nonlinear optimization

(General formulation of the problem; Important examples; Black box and iterative methods; Analytical and arithmetical complexity; Uniform grid method; Lower complexity bounds; Lower bounds for global optimization; Rules of the game.)

1.1.1 General formulation of the problem

Let us start by fixing the mathematical form of our main problem and the standard terminology. Let x be an n-dimensional real vector:

$$x = (x^{(1)}, \ldots, x^{(n)})^T \in R^n,$$

S be a subset of R^n, and $f_0(x), \ldots, f_m(x)$ be some real-valued functions of x. In the entire book we deal with different variants of the following general minimization problem:

$$\begin{aligned} &\min\ f_0(x), \\ \text{s.t.}\ \ & f_j(x)\ \&\ 0, \quad j = 1 \ldots m, \\ & x \in S, \end{aligned} \qquad (1.1.1)$$

where the sign & could be $\leq$, $\geq$ or $=$.

We call $f_0(x)$ the *objective* function of our problem, the vector function

$$f(x) = (f_1(x), \ldots, f_m(x))^T$$

is called the vector of *functional constraints*, the set S is called the *basic feasible set*, and the set

$$Q = \{x \in S \mid\ f_j(x) \leq 0,\ j = 1 \ldots m\}$$

is called the *feasible set* of problem (1.1.1). That is just a convention to consider a minimization problem. Instead, we could consider a maximization problem with the objective function $-f_0(x)$.

There is a natural classification of the *types* of minimization problems:

- *Constrained problems*: $Q \subset R^n$.
- *Unconstrained problems*: $Q \equiv R^n$.
- *Smooth problems*: all $f_j(x)$ are differentiable.
- *Nonsmooth problems*: there is a nondifferentiable component $f_k(x)$.
- *Linearly constrained problems*: all functional constraints are linear:

$$f_j(x) = \sum_{i=1}^{n} a_j^{(i)} x^{(i)} + b_j \equiv \langle a_j, x \rangle + b_j, \; j = 1 \dots m,$$

(here $\langle \cdot, \cdot \rangle$ stands for the *inner product* in R^n), and S is a polyhedron.

If $f_0(x)$ is also linear, then (1.1.1) is a *linear optimization problem*. If $f_0(x)$ is quadratic, then (1.1.1) is a *quadratic optimization problem*. If all f_j are quadratic, then this is a quadratically constrained quadratic problem.

There is also a classification based on the properties of feasible set.

- Problem (1.1.1) is called *feasible* if $Q \neq \emptyset$.
- Problem (1.1.1) is called *strictly feasible* if there exists $x \in \operatorname{int} Q$ such that $f_j(x) < 0$ (or > 0) for all inequality constraints and $f_j(x) = 0$ for all equality constraints. (*Slater condition.*)

Finally, we distinguish different types of solutions to (1.1.1):

- x^* is called the optimal *global solution* to (1.1.1) if $f_0(x^*) \leq f_0(x)$ for all $x \in Q$ (*global minimum*). In this case $f_0(x^*)$ is called the (global) *optimal value* of the problem.
- x^* is called a *local solution* to (1.1.1) if $f_0(x^*) \leq f_0(x)$ for all $x \in \operatorname{int} \bar{Q} \subset Q$ (*local minimum*).

Let us consider now several examples demonstrating the origin of the optimization problems.

EXAMPLE 1.1.1 Let $x^{(1)}, \dots, x^{(n)}$ be our *design variables*. Then we can fix some functional *characteristics* of our decision: $f_0(x), \dots, f_m(x)$. For example, we can consider a price of the project, amount of required resources, reliability of the system, etc. We fix the most important characteristics, $f_0(x)$, as our *objective*. For all others we impose some bounds: $a_j \leq f_j(x) \leq b_j$. Thus, we come up with the problem:

$$\begin{aligned} &\min\ f_0(x), \\ \text{s.t.:}\quad & a_j \leq f_j(x) \leq b_j,\ j = 1 \ldots m, \\ & x \in S, \end{aligned}$$

where S stands for the *structural* constraints, like nonnegativity or boundedness of some variables. □

EXAMPLE 1.1.2 Let our initial problem be as follows:

$$\text{Find } x \in R^n \text{ such that } f_j(x) = a_j,\ j = 1 \ldots m. \tag{1.1.2}$$

Then we can consider the problem:

$$\min_x \sum_{j=1}^m (f_j(x) - a_j)^2,$$

perhaps even with some additional constraints on x. If the optimal value of the latter problem is zero, we conclude that our initial problem (1.1.2) has a solution.

Note that the problem (1.1.2) is almost *universal*. It covers ordinary differential equations, partial differential equations, problems arising in Game Theory, and many others. □

EXAMPLE 1.1.3 Sometimes our decision variables $x^{(1)}, \ldots, x^{(n)}$ must be *integer*. This can be described by the following constraint:

$$\sin(\pi x^{(i)}) = 0, \quad i = 1 \ldots n.$$

Thus, we can also treat *integer optimization* problems:

$$\begin{aligned} &\min\ f_0(x), \\ \text{s.t.:}\quad & a_j \leq f_j(x) \leq b_j,\ j = 1 \ldots m, \\ & x \in S, \\ & \sin(\pi x^{(i)}) = 0,\ i = 1 \ldots n. \end{aligned}$$

□

Looking at these examples, a reader can understand the optimism of the pioneers of nonlinear optimization, which can be easily seen in the papers of the 1950's and 1960's. Our first impression should be, of course, as follows:

> Nonlinear optimization is a very important and promising application theory. It covers almost *all* needs of operations research and numerical analysis.

However, just by looking at the same examples, especially at Examples 1.1.2 and 1.1.3, a more suspicious (or more experienced) reader could come to the following conjecture:

> *In general, optimization problems are* unsolvable

Indeed, the real life is too complicated to believe in a universal tool, which can solve all problems at once.

However, such suspicions are not so important in science; that is a question of personal taste how much we trust them. Therefore it was definitely one of the most important events in optimization, when in the middle of the 1970s this conjecture was *proved* in some *strict* mathematical sense. The proof is so simple and remarkable, that we cannot avoid it in our course. But first of all, we should introduce a special language, which is necessary to speak about such things.

1.1.2 Performance of numerical methods

Let us imagine the following situation: We have a problem $\mathcal{P}$, which we are going to solve. We know that there are different numerical methods for doing so, and of course, we want to find a scheme that is the best for our $\mathcal{P}$. However, it turns out that we are looking for something that does not exist. In fact, maybe it does, but it is definitely not recommended to ask the winner for help. Indeed, consider a method for solving problem (1.1.1), which is doing nothing except reporting that $x^* = 0$. Of course, this method does not work for all problems *except* those for which the solution is indeed the origin. And for the latter problems the "performance" of such a scheme is unbeatable.

Thus, we cannot speak about the best method for a particular problem $\mathcal{P}$, but we can do so for a *class* of problems $\mathcal{F} \supset \mathcal{P}$. Indeed, usually the numerical methods are developed for solving many different problems with similar characteristics. Thus, a *performance* of method $\mathcal{M}$ on the whole class $\mathcal{F}$ is a natural characteristic of its efficiency.

Since we are going to speak about the performance of $\mathcal{M}$ on a class $\mathcal{F}$, we should assume that $\mathcal{M}$ does not have *complete* information about a particular problem $\mathcal{P}$.

> A *known* (for numerical scheme) "part" of problem $\mathcal{P}$ is called the *model* of the problem.

We denote the model by Σ. Usually the model consists of problem formulation, description of classes of functional components, etc.

In order to recognize the problem $\mathcal{P}$ (and solve it), the method should be able to collect specific information about $\mathcal{P}$. It is convenient to describe the process of collecting the data by the notion of an *oracle*. An oracle $\mathcal{O}$ is just a unit, which answers the successive questions of the method. The method $\mathcal{M}$ is trying to solve the problem $\mathcal{P}$ by collecting and handling the answers.

In general, each problem can be described by different models. Moreover, for each problem we can develop different types of oracles. But let us fix Σ and $\mathcal{O}$. In this case, it is natural to define the performance of $\mathcal{M}$ on $(\Sigma, \mathcal{O})$ as its performance on the *worst* $\mathcal{P}_{\rm w}$ from $(\Sigma, \mathcal{O})$. Note that this $\mathcal{P}_{\rm w}$ can be bad only for $\mathcal{M}$.

Further, what is the *performance* of $\mathcal{M}$ on $\mathcal{P}$? Let us start from the intuitive definition:

> Performance of $\mathcal{M}$ on $\mathcal{P}$ is the total amount of *computational efforts* that is required by method $\mathcal{M}$ to *solve the problem* $\mathcal{P}$.

In this definition there are two additional notions to be specified. First of all, what does it mean: to solve the problem? In some situations it could mean finding an *exact* solution. However, in many areas of numerical analysis that is impossible (and optimization is definitely such a case). Therefore,

> To solve the problem means to find an *approximate* solution to $\mathcal{M}$ with an accuracy $\epsilon > 0$.

The meaning of the words *with an accuracy* $\epsilon > 0$ is very important for our definitions. However, it is too early to speak about that now. We just introduce notation $\mathcal{T}_\epsilon$ for a stopping criterion; its meaning will be always precise for particular problem classes. Now we can give a formal definition of the problem class:

$$\mathcal{F} \equiv (\Sigma, \mathcal{O}, \mathcal{T}_\epsilon).$$

In order to solve a problem $\mathcal{P} \in \mathcal{F}$ we can apply an *iterative process*, which naturally describes any method $\mathcal{M}$ working with the oracle.

General Iterative Scheme.

Input: A starting point x_0 and an accuracy $\epsilon > 0$.
Initialization. Set $k = 0$, $I_{-1} = \emptyset$. Here k is iteration counter and I_k is accumulated *information set.*

(1.1.3)

Main loop:
1. Call oracle $\mathcal{O}$ at x_k.
2. Update the information set: $I_k = I_{k-1} \bigcup (x_k, \mathcal{O}(x_k))$.
3. Apply rules of method $\mathcal{M}$ to I_k and form point x_{k+1}.
4. Check criterion $\mathcal{T}_\epsilon$. If **yes** then form an output $\bar{x}$. Otherwise set $k := k+1$ and go to Step 1.

Now we can specify the term *computational efforts* in our definition of performance. In the scheme (1.1.3) we can easily find two most expensive steps. The first one is Step 1, where we call the oracle, and the second one is Step 3, where we form the next test point. Thus, we can introduce two measures of *complexity* of problem $\mathcal{P}$ for method $\mathcal{M}$:

Analytical complexity: The number of calls of oracle, which is required to solve problem $\mathcal{P}$ up to accuracy ϵ.

Arithmetical complexity: The total number of arithmetic operations (including the work of oracle and work of method), which is required to solve problem $\mathcal{P}$ up to accuracy ϵ.

Comparing the notions of analytical and arithmetical complexity, we can see that the second one is more realistic. However, for a particular method $\mathcal{M}$ as applied to a problem $\mathcal{P}$, the arithmetical complexity usually can be easily obtained from the analytical complexity and the complexity of the oracle. Therefore, in this course we will speak mainly about bounds on the analytical complexity for some problem classes.

There is one standard assumption on the oracle, which allows us to obtain the majority of the results on the analytical complexity of optimization problems. This assumption is called the ***local black box concept*** and it looks as follows:

Local black box
1. The only information available for the numerical scheme is the answer of the oracle. 2. The oracle is *local*: A small variation of the problem far enough from the test point x does not change the answer at x.

This concept is very useful in numerical analysis. Of course, its first part looks like an artificial wall between the method and the oracle. It seems natural to give the method an access to internal structure of the problem. However, we will see that for problems with rather complicated structure this access is almost useless. For more simple problems it could help. We will see that in the last chapter of the book.

To conclude the section, let us mention that the standard formulation (1.1.1) is called a *functional model* of optimization problems. Usually, for such models the standard assumptions are related to the smoothness of functional components. In accordance to degree of smoothness we can apply different types of oracle:

- *Zero-order* oracle: returns the value $f(x)$.
- *First-order* oracle: returns $f(x)$ and the gradient $f'(x)$.
- *Second-order* oracle: returns $f(x)$, $f'(x)$ and the Hessian $f''(x)$.

1.1.3 Complexity bounds for global optimization

Let us try to apply the formal language, introduced in the previous section, to a particular problem class. Consider, for example, the following problem:

$$\min_{x \in B_n} f(x). \tag{1.1.4}$$

In our terminology, this is a constrained minimization problem without functional constraints. The basic feasible set of this problem is B_n, an

n-dimensional box in R^n:

$$B_n = \{x \in R^n \mid 0 \le x^{(i)} \le 1, \ i = 1 \dots n\}.$$

Let us measure distances in R^n using l_∞-norm:

$$\|x\|_\infty = \max_{1 \le i \le n} |x^{(i)}|.$$

Assume that, with respect to this norm,

> the objective function $f(x)$ is *Lipschitz continuous* on B_n:
>
> $$| f(x) - f(y) | \le L \| x - y \|_\infty \quad \forall x, y \in B_n, \tag{1.1.5}$$
>
> with some constant L (*Lipschitz constant*).

Consider a very simple method for solving (1.1.4), which is called the *uniform grid method.* This method $\mathcal{G}(p)$ has one integer input parameter p. Its scheme is as follows.

Method $\mathcal{G}(p)$
1. Form $(p+1)^n$ points $x_{(i_1,\dots,i_n)} = \left(\frac{i_1}{p}, \frac{i_2}{p}, \dots, \frac{i_n}{p}\right)^T,$ where $(i_1, \dots, i_n) \in \{0, \dots, p\}^n$. (1.1.6)
2. Among all points $x_{(i_1,\dots,i_n)}$ find a point $\bar{x}$, which has the minimal value of objective function.
3. Return the pair $(\bar{x}, f(\bar{x}))$ as a result.

Thus, this method forms a uniform grid of the test points inside the box B_n, computes the minimum value of the objective over this grid and returns this value as an approximate solution to problem (1.1.4). In our terminology, this is a zero-order iterative method without any influence of the accumulated information on the sequence of test points. Let us find its efficiency estimate.

THEOREM 1.1.1 *Let f^* be the global optimal value of problem (1.1.4). Then*

$$f(\bar{x}) - f^* \leq \frac{L}{2p}.$$

Proof: Let x_* be a global minimum of our problem. Then there exist coordinates $(i_1, i_2, \ldots, i_n)$ such that

$$x \equiv x_{(i_1, i_2, \ldots, i_n)} \leq x^* \leq x_{(i_1+1, i_2+1, \ldots, i_n+1)} \equiv y$$

(here and in the sequel we write $x \leq y$ for $x, y \in R^n$ if and only if $x^{(i)} \leq y^{(i)}$ for all $i = 1 \ldots n$). Note that $y^{(i)} - x^{(i)} = \frac{1}{p}$ for $i = 1 \ldots n$, and

$$x_*^{(i)} \in [x^{(i)}, y^{(i)}], \quad i = 1 \ldots n.$$

Denote $\hat{x} = (x + y)/2$. Let us form a point $\tilde{x}$ as follows:

$$\tilde{x}^{(i)} = \begin{cases} y^{(i)}, & \text{if } x_*^{(i)} \geq \hat{x}^{(i)}, \\ x^{(i)}, & \text{otherwise.} \end{cases}$$

It is clear that $\mid \tilde{x}^{(i)} - x_*^{(i)} \mid \leq \frac{1}{2p}$, $i = 1 \ldots n$. Therefore

$$\| \tilde{x} - x^* \|_\infty = \max_{1 \leq i \leq n} |\tilde{x}^{(i)} - x_*^{(i)}| \leq \frac{1}{2p}.$$

Since $\tilde{x}$ belongs to our grid, we conclude that

$$f(\bar{x}) - f(x_*) \leq f(\tilde{x}) - f(x_*) \leq L \| \tilde{x} - x_* \|_\infty \leq \frac{L}{2p}.$$

□

Let us finish the definition of our problem class. Define our goal as follows:

$$\text{Find } \bar{x} \in B_n : f(\bar{x}) - f^* \leq \epsilon. \tag{1.1.7}$$

Then we immediately get the following result.

COROLLARY 1.1.1 *Analytical complexity of the problem class (1.1.4), (1.1.5), (1.1.7) for method $\mathcal{G}$ is at most*

$$\mathcal{A}(\mathcal{G}) = \left(\lfloor \tfrac{L}{2\epsilon} \rfloor + 2 \right)^n,$$

(here $\lfloor a \rfloor$ is an integer part of a).

Proof: Take $p = \lfloor \frac{L}{2\epsilon} \rfloor + 1$. Then $p \geq \frac{L}{2\epsilon}$, and, in view of Theorem 1.1.1, we have $f(\bar{x}) - f^* \leq \frac{L}{2p} \leq \epsilon$. Note that we constrict $(p+1)^n$ points. □

Thus, $\mathcal{A}(\mathcal{G})$ justifies an *upper* complexity bound for our problem class.

This result is quite informative, but we still have some questions. Firstly, it may happen that our proof is too rough and the real performance of $\mathcal{G}(p)$ is much better. Secondly, we still cannot be sure that $\mathcal{G}(p)$ is a reasonable method for solving (1.1.4). There may exist other schemes with much higher performance.

In order to answer these questions, we need to derive *lower complexity bounds* for the problem class (1.1.4), (1.1.5), (1.1.7). The main features of such bounds are as follows.

- They are based on the *black box* concept.
- These bounds are valid for all reasonable iterative schemes. Thus, they provide us with a lower estimate for *analytical complexity* on the problem class.
- Very often such bounds employ the idea of the *resisting* oracle.

For us only the notion of the resisting oracle is new. Therefore, let us discuss it in more detail.

A resisting oracle tries to create a *worst* problem for each particular method. It starts from an "empty" function and it tries to answer each call of the method in the worst possible way. However, the answers must be *compatible* with the previous answers and with the description of the problem class. Then, after termination of the method it is possible to *reconstruct* a problem, which fits completely the final information set accumulated by the algorithm. Moreover, if we launch this method on this problem, it will reproduce the same sequence of the test points since it will have the same sequence of answers from the oracle.

Let us show how that works for problem (1.1.4). Consider the class of problems $\mathcal{C}$ defined as follows:

Model:	$\min\limits_{x \in B_n} f(x)$, $f(x)$ is l_∞-Lipschitz continuous on B_n.
Oracle:	Zero-order local black box.
Approximate solution:	Find $\bar{x} \in B_n : f(\bar{x}) - f^* \leq \epsilon$.

THEOREM 1.1.2 *For $\epsilon < \frac{1}{2}L$ the analytical complexity of $\mathcal{C}$ for zero-order methods is at least* $\left(\lfloor \frac{L}{2\epsilon} \rfloor \right)^n$.

Proof: Denote $p = \lfloor \frac{L}{2\epsilon} \rfloor$ (≥ 1). Assume that there exists a method, which needs $N < p^n$ calls of oracle to solve any problem from $\mathcal{C}$. Let us apply this method to the following resisting strategy:

Oracle returns $f(x) = 0$ at any test point x.

Therefore this method can find only $\bar{x} \in B_n$ with $f(\bar{x}) = 0$. However, note that there exists $\hat{x} \in B_n$ such that

$$\hat{x} + \frac{1}{p}e \in B_n, \quad e = (1, \dots, 1)^T \in R^n,$$

and there were no test points inside the box $B = \{x \mid \hat{x} \leq x \leq \hat{x} + \frac{1}{p}e\}$. Denote $x_* = \hat{x} + \frac{1}{2p}e$ and consider the function

$$\bar{f}(x) = \min\{0, L \parallel x - x_* \parallel_\infty - \epsilon\},$$

Clearly, this function is l_∞-Lipschitz continuous with the constant L and its global optimal value is $-\epsilon$. Moreover, $\bar{f}(x)$ differs from zero only inside the box $B' = \{x \mid \parallel x - x_* \parallel_\infty \leq \frac{\epsilon}{L}\}$. Since $2p \leq L/\epsilon$, we conclude that

$$B' \subseteq B \equiv \{x \mid \parallel x - \tilde{x} \parallel_\infty \leq \frac{1}{2p}\}.$$

Thus, $\bar{f}(x)$ is equal to zero *at all test points* of our method. Since the accuracy of the result of our method is ϵ, we come to the following conclusion: If the number of calls of the oracle is less than p^n, then the accuracy of the result cannot be better than ϵ. □

Now we can say much more about the performance of the uniform grid method. Let us compare its efficiency estimate with the lower bound:

$$\mathcal{G}: \left(\lfloor \frac{L}{2\epsilon} \rfloor + 2\right)^n, \quad \text{Lower bound: } \left(\lfloor \frac{L}{2\epsilon} \rfloor\right)^n.$$

Thus, if $\epsilon = O(\frac{L}{n})$, the lower and upper bounds coincide up to a constant multiplicative factor. This implies that $\mathcal{G}(p)$ is an *optimal* method for $\mathcal{C}$.

At the same time, Theorem 1.1.2 supports our initial claim that the general optimization problems are unsolvable. Let us look at the following example.

EXAMPLE 1.1.4 Consider the problem class $\mathcal{F}$ defined by the following parameters:

$$L = 2, \quad n = 10, \quad \epsilon = 0.01.$$

Note that the size of the problem is very small and we ask only for 1% accuracy.

The lower complexity bound for this class is $\left(\frac{L}{2\epsilon}\right)^n$. Let us compute it for our example.

Lower bound:	10^{20} calls of oracle
Complexity of oracle:	at least n arithmetic operations (a.o.)
Total complexity:	10^{21} a.o.
Work station:	10^6 a.o. per second
Total time:	10^{15} seconds
One year:	less than $3.2 \cdot 10^7$ sec.
We need:	31 250 000 years.

This estimate is so disappointing that we cannot keep any hope that such problems may become solvable in a future. Let us just play with the parameters of the problem class.

- If we change n to $n+1$, then the estimate is multiplied by one hundred. Thus, for $n = 11$ our lower bound is valid for a much more powerful computer.

- On the contrary, if we multiply ϵ by two, we reduce the complexity by a factor of a thousand. For example, if $\epsilon = 8\%$, then we need only two weeks. □

We should note, that the lower complexity bounds for problems with smooth functions, or for high-order methods are not much better than those of Theorem 1.1.2. This can be proved using the same arguments and we leave the proof as an exercise for the reader. Comparison of the above results with the *upper* bounds for NP-hard problems, which are considered as a classical example of very difficult problems in combinatorial optimization, is also quite disappointing. Hard combinatorial problems need 2^n a.o. only!

To conclude this section, let us compare our situation with one in some other fields of numerical analysis. It is well known, that the uniform grid approach is a standard tool in many domains. For example, if we need to compute numerically the value of the integral of a univariate function

$$\mathcal{I} = \int_0^1 f(x)dx,$$

the standard way to proceed is to form a discrete sum

$$S_n = \frac{1}{N}\sum_{i=1}^{n} f(x_i), \quad x_i = \frac{i}{N}, \ i = 1 \dots N.$$

If $f(x)$ is Lipschitz continuous, then this value can be used as an approximation to $\mathcal{I}$:

$$N = L/\epsilon \quad \Rightarrow \quad | \mathcal{I} - S_N | \leq \epsilon.$$

Note that in our terminology this is exactly the uniform grid approach. Moreover, that is a standard way for approximating the integrals. The reason why it works here lies in the *dimension* of problems. For integration the standard dimensions are very small (up to three), and in optimization sometimes we need to solve problems with several millions of variables.

1.1.4 Identity cards of the fields

After the pessimistic results of the previous section, first of all we should understand what could be our goal in theoretical analysis of optimization problems. It seems, everything is clear for general global optimization. But maybe the goals of this field are too ambitious? Maybe in some practical problems we would be satisfied by much less "optimal" solution? Or, maybe there are some interesting problem classes, which are not so dangerous as the class of general continuous functions?

In fact, each of these questions can be answered in a different way. And this way defines the style of research (or rules of the game) in the different fields of nonlinear optimization. If we try to classify these fields, we can easily see that they differ one from another in the following aspects:

- Goals of the methods.
- Classes of functional components.
- Description of the oracle.

These aspects define in a natural way the list of desired properties of the optimization methods. Let us present the "identity cards" of the fields, which we are going to consider in the book.

Name: General global optimization. (Section 1.1)
Goals: Find a global minimum.
Functional class: Continuous functions.
Oracle: $0-1-2$ order black box.
Desired properties: Convergence to a global minimum.
Features: From theoretical point of view, this game is too short. We always lose it.
Problem sizes: There are examples of solving problems with thousands of variables. However, no guarantee for success even for very small problems.
History: Starts from 1955. Several local peaks of interest related to new heuristic ideas (simulated annealing, neural networks, genetic algorithms).

Name: Nonlinear optimization. (Sections 1.2, 1.3)
Goals: Find a local minimum.
Functional class: Differentiable functions.
Oracle: $1-2$ order black box.
Desired properties: Convergence to a local minimum. Fast convergence.
Features: Variability of approaches. Most widespread software. The goal is not always acceptable and reachable.
Problem sizes: up to 1000 variables.
History: Starts from 1955. Peak period: 1965 - 1985. Theoretical activity now is rather low.

Name: Convex optimization. (Chapters 2, 3)
Goals: Find a global minimum.
Functional class: Convex sets and functions.
Oracle: 1st-order black box.
Desired properties: Convergence to a global minimum. Rate of convergence depends on the dimension.
Features: Very rich and interesting theory. Comprehensive complexity theory. Efficient practical methods. The problem class is sometimes restrictive.
Problem sizes: up to 1000 variables.
History: Starts from 1970. Peak period: 1975 - 1985 (terminated by explosion of interior-point ideas). Theoretical activity now is growing up.

Name: Interior-point polynomial-time methods. (Chapter 4)
Goals: Find a global minimum.
Functional class: Convex sets and functions with explicit structure.
Oracle: 2nd-order black box oracle, which is not local.
Desired properties: Fast convergence to a global minimum. Rate of convergence depends on the structure of the problem.
Features: Very new and perspective theory. Avoid the black box concept. The problem class is practically the same as in convex optimization.
Problem sizes: Sometimes up to 10 000 000 variables.
History: Starts from 1984. Peak period: 1990 – .... Very high theoretical activity just now.

1.2 Local methods in unconstrained minimization

(Relaxation and approximation; Necessary optimality conditions; Sufficient optimality conditions; Class of differentiable functions; Class of twice differentiable functions; Gradient method; Rate of convergence; Newton method.)

1.2.1 Relaxation and approximation

The simplest goal of general nonlinear optimization is to find a local minimum of a differentiable function. In general, the global structure of such a function is not simpler than that one of a Lipschitz continuous function. Therefore, even for reaching such a restricted goal, it is necessary to follow some special principles, which guarantee the convergence of a minimization process.

The majority of general nonlinear optimization methods are based on the idea of *relaxation:*

We call the sequence $\{a_k\}_{k=0}^{\infty}$ a *relaxation sequence* if

$$a_{k+1} \leq a_k \quad \forall k \geq 0.$$

In this section we consider several methods for solving the following unconstrained minimization problem

$$\min_{x \in R^n} f(x), \tag{1.2.1}$$

where $f(x)$ is a smooth function. In order to do so, we generate a relaxation sequence $\{f(x_k)\}_{k=0}^{\infty}$:

$$f(x_{k+1}) \leq f(x_k), \quad k = 0, 1, \ldots \quad .$$

This strategy has the following important advantages:

1. If $f(x)$ is bounded below on R^n, then the sequence $\{f(x_k)\}_{k=0}^{\infty}$ converges.

2. In any case we improve the initial value of the objective function.

However, it would be impossible to implement the idea of relaxation without employing another fundamental principle of numerical analysis, the *approximation*. In general,

> To approximate means to replace an initial complex object by a simplified one, which is close by its properties to the original.

In nonlinear optimization we usually apply *local* approximations based on derivatives of nonlinear functions. These are the first- and the second-order approximations (or, the linear and quadratic approximations).

Let $f(x)$ be differentiable at $\bar{x}$. Then for $y \in R^n$ we have

$$f(y) = f(\bar{x}) + \langle f'(\bar{x}), y - \bar{x} \rangle + o(\| y - \bar{x} \|),$$

where $o(r)$ is some function of $r \geq 0$ such that

$$\lim_{r \downarrow 0} \frac{1}{r} o(r) = 0, \quad o(0) = 0.$$

In the sequel we fix the notation $\| \cdot \|$ for the standard *Euclidean* norm in R^n:

$$\|x\| = \left[\sum_{i=1}^{n} \left(x^{(i)} \right)^2 \right]^{1/2}.$$

The linear function $f(\bar{x}) + \langle f'(\bar{x}), y - \bar{x} \rangle$ is called the *linear approximation* of f at $\bar{x}$. Recall that the vector $f'(x)$ is called the *gradient* of function f at x. Considering the points $y_i = \bar{x} + \epsilon e_i$, where e_i is the ith coordinate vector in R^n, and taking the limit in $\epsilon \to 0$, we obtain the following coordinate representation of the gradient:

$$f'(x) = \left(\frac{\partial f(x)}{\partial x^{(1)}}, \ldots, \frac{\partial f(x)}{\partial x^{(n)}} \right)^T.$$

Let us mention two important properties of the gradient. Denote by $\mathcal{L}_f(\alpha)$ the *level set* of $f(x)$:

$$\mathcal{L}_f(\alpha) = \{x \in R^n \mid f(x) \leq \alpha\}.$$

Consider the set of directions that are *tangent* to $\mathcal{L}_f(f(\bar{x}))$ at $\bar{x}$:

$$S_f(\bar{x}) = \left\{ s \in R^n \mid s = \lim_{\substack{y_k \to \bar{x}, \\ f(y_k)=f(\bar{x})}} \frac{y_k - \bar{x}}{\| y_k - \bar{x} \|} \right\}.$$

LEMMA 1.2.1 *If $s \in S_f(\bar{x})$, then $\langle f'(\bar{x}), s\rangle = 0$.*

Proof: Since $f(y_k) = f(\bar{x})$, we have

$$f(y_k) = f(\bar{x}) + \langle f'(\bar{x}), y_k - \bar{x}\rangle + o(\| y_k - \bar{x} \|) = f(\bar{x}).$$

Therefore $\langle f'(\bar{x}), y_k - \bar{x}\rangle + o(\| y_k - \bar{x} \|) = 0$. Dividing this equation by $\| y_k - \bar{x} \|$ and taking the limit in $y_k \to \bar{x}$, we obtain the result. □

Let s be a direction in R^n, $\| s \|= 1$. Consider the local decrease of $f(x)$ along s:

$$\Delta(s) = \lim_{\alpha \downarrow 0} \frac{1}{\alpha}[f(\bar{x} + \alpha s) - f(\bar{x})].$$

Note that $f(\bar{x} + \alpha s) - f(\bar{x}) = \alpha\langle f'(\bar{x}), s\rangle + o(\alpha)$. Therefore

$$\Delta(s) = \langle f'(\bar{x}), s\rangle.$$

Using the Cauchy–Schwartz inequality:

$$- \| x \| \cdot \| y \| \leq \langle x, y\rangle \leq \| x \| \cdot \| y \|,$$

we obtain $\Delta(s) = \langle f'(\bar{x}), s\rangle \geq - \| f'(\bar{x}) \|$. Let us take

$$\bar{s} = -f'(\bar{x}) / \| f'(\bar{x} \| .$$

Then

$$\Delta(\bar{s}) = -\langle f'(\bar{x}), f'(\bar{x})\rangle / \| f'(\bar{x}) \| = - \| f'(\bar{x}) \| .$$

Thus, the direction $-f'(\bar{x})$ (the *antigradient*) is the direction of the *fastest local decrease* of $f(x)$ at point $\bar{x}$.

The next statement is probably the most fundamental fact in optimization.

THEOREM 1.2.1 (First-order optimality condition.)
Let x^* be a local minimum of differentiable function $f(x)$. Then

$$f'(x^*) = 0.$$

Proof: Since x^* is a local minimum of $f(x)$, then there exists $r > 0$ such that for all y, $\|y - x^*\| \leq r$, we have $f(y) \geq f(x^*)$. Since f is differentiable, this implies that

$$f(y) = f(x^*) + \langle f'(x^*), y - x^* \rangle + o(\| y - x^* \|) \geq f(x^*).$$

Thus, for all s, $\| s \| = 1$, we have $\langle f'(x^*), s \rangle \geq 0$. Consider the directions s and $-s$; we get

$$\langle f'(x^*), s \rangle = 0, \quad \forall s, \ \| s \| = 1.$$

Finally, choosing $s = e_i$, $i = 1 \dots n$, where e_i is the ith coordinate vector in R^n, we obtain $f'(x^*) = 0$. □

COROLLARY 1.2.1 *Let x^* be a local minimum of a differentiable function $f(x)$ subject to linear equality constraints*

$$x \in \mathcal{L} \equiv \{x \in R^n \mid Ax = b\} \neq \emptyset,$$

where A is an $m \times n$-matrix and $b \in R^m$, $m < n$. Then there exists a vector of multipliers λ^* such that

$$f'(x^*) = A^T \lambda^*. \qquad (1.2.2)$$

Proof: Consider some vectors u_i, $i = 1 \dots k$, that form a basis of the null space of matrix A. Then any $x \in \mathcal{L}$ can be represented as follows:

$$x = x(y) \equiv x^* + \sum_{i=1}^{k} y^{(i)} u_i, \quad y \in R^k.$$

Moreover, the point $y = 0$ is a local minimum of the function $\phi(y) = f(x(y))$. In view of Theorem 1.2.1, $\phi'(0) = 0$. This means that

$$\frac{\partial \phi(0)}{\partial y^{(i)}} = \langle f'(x^*), u_i \rangle = 0, \quad i = 1 \dots k,$$

and (1.2.2) follows. □

Note that we have proved only a *necessary* condition of a local minimum. The points satisfying this condition are called the *stationary points* of function f. In order to see that such points are not always the local minima, it is enough to look at function $f(x) = x^3$, $x \in R^1$, at $x = 0$.

Let us introduce now the second-order approximation. Let function $f(x)$ be twice differentiable at $\bar{x}$. Then

$$f(y) = f(\bar{x}) + \langle f'(\bar{x}), y - \bar{x}\rangle + \tfrac{1}{2}\langle f''(\bar{x})(y-\bar{x}), y - \bar{x}\rangle + o(\| y - \bar{x} \|^2).$$

The quadratic function

$$f(\bar{x}) + \langle f'(\bar{x}), y - \bar{x}\rangle + \tfrac{1}{2}\langle f''(\bar{x})(y-\bar{x}), y - \bar{x}\rangle$$

is called the *quadratic* (or *second-order)* approximation of function f at $\bar{x}$. Recall that the $(n \times n)$-matrix $f''(x)$ has the following entries:

$$(f''(x))^{(i,j)} = \frac{\partial^2 f(x)}{\partial x^{(i)} \partial x^{(j)}}.$$

It is called the *Hessian* of function f at x. Note that the Hessian is a symmetric matrix:

$$f''(x) = [f''(x)]^T.$$

The Hessian can be seen as a derivative of the vector function $f'(x)$:

$$f'(y) = f'(\bar{x}) + f''(\bar{x})(y - \bar{x}) + \mathbf{o}(\| y - \bar{x} \|),$$

where $\mathbf{o}(r)$ is a vector function such that $\lim\limits_{r \downarrow 0} \frac{1}{r} \| \mathbf{o}(r) \| = 0$ and $\mathbf{o}(0) = 0$.

Using the second–order approximation, we can write down the second–order optimality conditions. In what follows notation $A \succeq 0$, used for a symmetric matrix A, means that A is *positive semidefinite*:

$$\langle Ax, x\rangle \geq 0 \quad \forall x \in R^n.$$

Notation $A \succ 0$ means that A is *positive definite* (above inequality must be strict for $x \neq 0$).

THEOREM 1.2.2 (Second-order optimality condition.)
Let x^* be a local minimum of twice differentiable function $f(x)$. Then

$$f'(x^*) = 0, \qquad f''(x^*) \succeq 0.$$

Proof: Since x^* is a local minimum of function $f(x)$, there exists $r > 0$ such that for all y, $\|y - x^*\| \leq r$, we have

$$f(y) \geq f(x^*).$$

In view of Theorem 1.2.1, $f'(x^*) = 0$. Therefore, for any such y,

$$f(y) = f(x^*) + \langle f''(x^*)(y - x^*), y - x^*\rangle + o(\| y - x^* \|^2) \geq f(x^*).$$

Thus, $\langle f''(x^*)s, s\rangle \geq 0$, for all s, $\| s \| = 1$. □

Again, the above theorem is a *necessary* (second-order) characteristic of a local minimum. Let us prove a sufficient condition.

THEOREM 1.2.3 *Let function $f(x)$ be twice differentiable on R^n and let x^* satisfy the following conditions:*

$$f'(x^*) = 0, \quad f''(x^*) \succ 0.$$

Then x^* is a strict local minimum of $f(x)$.

(Sometimes, instead of *strict* we say *isolated.*)
Proof: Note that in a small neighborhood of point x^* function $f(x)$ can be represented as

$$f(y) = f(x^*) + \tfrac{1}{2}\langle f''(x^*)(y - x^*), y - x^*\rangle + o(\| y - x^* \|^2).$$

Since $\frac{o(r)}{r} \to 0$, there exists a value $\bar{r}$ such that for all $r \in [0, \bar{r}]$ we have

$$| o(r) | \leq \tfrac{r}{4}\lambda_1(f''(x^*)),$$

where $\lambda_1(f''(x^*))$ is the smallest eigenvalue of matrix $f''(x^*)$. Recall, that in view of our assumption, this eigenvalue is positive. Therefore, for any y, $\| y - x^*\| \leq \bar{r}$ we have

$$\begin{aligned} f(y) \geq\ & f(x^*) + \tfrac{1}{2}\lambda_1(f''(x^*)) \| y - x^* \|^2 + o(\| y - x^* \|^2) \\ \geq\ & f(x^*) + \tfrac{1}{4}\lambda_1(f''(x^*)) \| y - x^* \|^2 > f(x^*). \end{aligned}$$

□

1.2.2 Classes of differentiable functions

It is well known that any continuous function can be approximated by a smooth function with arbitrarily small accuracy. Therefore, assuming only differentiability of the objective function we cannot get any reasonable properties of minimization processes. Hence, we have to impose some additional assumptions on the magnitude of the derivatives. Traditionally, in optimization such assumptions are presented in the form of a *Lipschitz condition* for a derivative of certain order.

Let Q be a subset of R^n. We denote by $C_L^{k,p}(Q)$ the class of functions with the following properties:

- any $f \in C_L^{k,p}(Q)$ is k times continuously differentiable on Q.
- Its pth derivative is Lipschitz continuous on Q with the constant L:

$$\| f^{(p)}(x) - f^{(p)}(y) \| \leq L \| x - y \|$$

for all $x, y \in Q$.

Clearly, we always have $p \leq k$. If $q \geq k$, then $C_L^{q,p}(Q) \subseteq C_L^{k,p}(Q)$. For example, $C_L^{2,1}(Q) \subseteq C_L^{1,1}(Q)$. Note also that these classes possess the following property: *if* $f_1 \in C_{L_1}^{k,p}(Q)$, $f_2 \in C_{L_2}^{k,p}(Q)$ *and* $\alpha, \beta \in R^1$, *then for*

$$L_3 = \mid \alpha \mid L_1 + \mid \beta \mid L_2$$

we have $\alpha f_1 + \beta f_2 \in C_{L_3}^{k,p}(Q)$.

We use notation $f \in C^k(Q)$ for a function f which is k times continuously differentiable on Q.

For us the most important class of functions of the above type will be $C_L^{1,1}(R^n)$, the class of functions with Lipschitz continuous gradient. By definition, the inclusion $f \in C_L^{1,1}(R^n)$ implies that

$$\| f'(x) - f'(y) \| \leq L \| x - y \| \tag{1.2.3}$$

for all $x, y \in R^n$. Let us give a sufficient condition for that inclusion.

LEMMA 1.2.2 *Function* $f(x)$ *belongs to* $C_L^{2,1}(R^n) \subset C_L^{1,1}(R^n)$ *if and only if*

$$\| f''(x) \| \leq L, \quad \forall x \in R^n. \tag{1.2.4}$$

Proof. Indeed, for any $x, y \in R^n$ we have

$$\begin{aligned} f'(y) &= f'(x) + \int_0^1 f''(x + \tau(y - x))(y - x) d\tau \\ &= f'(x) + \left(\int_0^1 f''(x + \tau(y - x)) d\tau \right) \cdot (y - x). \end{aligned}$$

Therefore, if condition (1.2.4) is satisfied then

$$\begin{aligned} \| f'(y) - f'(x) \| &= \left\| \left(\int_0^1 f''(x + \tau(y - x)) d\tau \right) \cdot (y - x) \right\| \\ &\leq \left\| \int_0^1 f''(x + \tau(y - x)) d\tau \right\| \cdot \| y - x \| \\ &\leq \int_0^1 \| f''(x + \tau(y - x)) \| d\tau \cdot \| y - x \| \\ &\leq L \| y - x \| . \end{aligned}$$

On the other hand, if $f \in C_L^{2,1}(R^n)$, then for any $s \in R^n$ and $\alpha > 0$, we have

$$\left\| \left(\int_0^\alpha f''(x + \tau s) d\tau \right) \cdot s \right\| = \| f'(x + \alpha s) - f'(x) \| \leq \alpha L \| s \| .$$

Dividing this inequality by α and tending $\alpha \downarrow 0$, we obtain (1.2.4). □

This simple result provides us with many examples of functions with Lipschitz continuous gradient.

EXAMPLE 1.2.1 1. Linear function $f(x) = \alpha + \langle a, x \rangle \in C_0^{1,1}(R^n)$ since

$$f'(x) = a, \quad f''(x) = 0.$$

2. For the quadratic function $f(x) = \alpha + \langle a, x \rangle + \frac{1}{2}\langle Ax, x \rangle$ with $A = A^T$ we have

$$f'(x) = a + Ax, \quad f''(x) = A.$$

Therefore $f(x) \in C_L^{1,1}(R^n)$ with $L = \| A \|$.

3. Consider the function of one variable $f(x) = \sqrt{1 + x^2}$, $x \in R^1$. We have

$$f'(x) = \frac{x}{\sqrt{1 + x^2}}, \quad f''(x) = \frac{1}{(1 + x^2)^{3/2}} \leq 1.$$

Therefore $f(x) \in C_1^{1,1}(R)$. □

The next statement is important for the geometric interpretation of functions from $C_L^{1,1}(R^n)$

LEMMA 1.2.3 *Let $f \in C_L^{1,1}(R^n)$. Then for any x, y from R^n we have*

$$| f(y) - f(x) - \langle f'(x), y - x \rangle | \leq \frac{L}{2} \| y - x \|^2 . \qquad (1.2.5)$$

Proof: For all x, $y \in R^n$ we have

$$\begin{aligned} f(y) &= f(x) + \int_0^1 \langle f'(x + \tau(y - x)), y - x \rangle d\tau \\ &= f(x) + \langle f'(x), y - x \rangle + \int_0^1 \langle f'(x + \tau(y - x)) - f'(x), y - x \rangle d\tau. \end{aligned}$$

Therefore

$$
\begin{aligned}
&| f(y) - f(x) - \langle f'(x), y - x\rangle | \\
&= | \int_0^1 \langle f'(x + \tau(y-x)) - f'(x), y - x\rangle d\tau | \\
&\leq \int_0^1 | \langle f'(x + \tau(y-x)) - f'(x), y - x\rangle | d\tau \\
&\leq \int_0^1 \| f'(x + \tau(y-x)) - f'(x) \| \cdot \| y - x \| d\tau \\
&\leq \int_0^1 \tau L \| y - x \|^2 d\tau = \tfrac{L}{2} \| y - x \|^2 .
\end{aligned}
$$

□

Geometrically, we can draw the following picture. Consider a function f from $C_L^{1,1}(R^n)$. Let us fix some $x_0 \in R^n$ and define two quadratic functions

$$
\begin{aligned}
\phi_1(x) &= f(x_0) + \langle f'(x_0), x - x_0\rangle + \tfrac{L}{2} \| x - x_0 \|^2, \\
\phi_2(x) &= f(x_0) + \langle f'(x_0), x - x_0\rangle - \tfrac{L}{2} \| x - x_0 \|^2 .
\end{aligned}
$$

Then *graph* of function f is located between the graphs of ϕ_1 and ϕ_2:

$$\phi_1(x) \geq f(x) \geq \phi_2(x), \quad \forall x \in R^n.$$

Let us prove a similar result for the class of twice differentiable functions. Our main class of functions of that type will be $C_M^{2,2}(R^n)$, the class of twice differentiable functions with Lipschitz continuous Hessian. Recall that for $f \in C_M^{2,2}(R^n)$ we have

$$\| f''(x) - f''(y) \| \leq M \| x - y \| \tag{1.2.6}$$

for all x, $y \in R^n$.

LEMMA 1.2.4 *Let $f \in C_L^{2,2}(R^n)$. Then for any x, y from R^n we have*

$$\| f'(y) - f'(x) - f''(x)(y-x) \| \leq \tfrac{M}{2} \| y - x \|^2, \tag{1.2.7}$$

$$
\begin{aligned}
&|f(y) - f(x) - \langle f'(x), y - x\rangle - \tfrac{1}{2}\langle f''(x)(y-x), y - x\rangle| \\
&\quad \leq \tfrac{M}{6} \| y - x \|^3 .
\end{aligned}
\tag{1.2.8}
$$

Proof: Let us fix some x, $y \in R^n$. Then

$$f'(y) = f'(x) + \int_0^1 f''(x + \tau(y - x))(y - x)d\tau$$

$$= f'(x) + f''(x)(y - x) + \int_0^1 (f''(x + \tau(y - x)) - f''(x))(y - x)d\tau.$$

Therefore

$$\begin{aligned} & \| f'(y) - f'(x) - f''(x)(y - x) \| \\ = \; & \| \int_0^1 (f''(x + \tau(y - x)) - f''(x))(y - x)d\tau \| \\ \leq \; & \int_0^1 \| (f''(x + \tau(y - x)) - f''(x))(y - x) \| d\tau \\ \leq \; & \int_0^1 \| f''(x + \tau(y - x)) - f''(x) \| \cdot \| y - x \| d\tau \\ \leq \; & \int_0^1 \tau M \| y - x \|^2 d\tau = \frac{M}{2} \| y - x \|^2 . \end{aligned}$$

Inequality (1.2.8) can be proved in a similar way. □

COROLLARY 1.2.2 *Let $f \in C_M^{2,2}(R^n)$ and $\| y - x \| = r$. Then*

$$f''(x) - MrI_n \preceq f''(y) \preceq f''(x) + MrI_n,$$

where I_n is the unit matrix in R^n.

(Recall that for matrices A and B we write $A \succeq B$ if $A - B \succeq 0$.)
Proof: Denote $G = f''(y) - f''(x)$. Since $f \in C_M^{2,2}(R^n)$, we have $\| G \| \leq Mr$. This means that eigenvalues of the symmetric matrix G, $\lambda_i(G)$, satisfy the following inequality:

$$| \lambda_i(G) | \leq Mr, \quad i = 1 \ldots n.$$

Hence, $-MrI_n \preceq G \equiv f''(y) - f''(x) \preceq MrI_n$. □

1.2.3 Gradient method

Now we are completely ready for studying the convergence rate of unconstrained minimization methods. Let us start from the simplest scheme. We already know that antigradient is a direction of locally steepest descent of differentiable function. Since we are going to find its local minimum, the following scheme is the first to be tried:

Gradient method
Choose $x_0 \in R^n$.
Iterate $x_{k+1} = x_k - h_k f'(x_k)$, $k = 0, 1, \ldots$.

(1.2.9)

We will refer to this scheme as a *gradient method*. The scalar factor of the gradient, h_k, is called the *step size*. Of course, it must be positive.

There are many variants of this method, which differ one from another by the *step-size strategy*. Let us consider the most important examples.

1. The sequence $\{h_k\}_{k=0}^{\infty}$ is chosen *in advance*. For example,

$$h_k = h > 0, \quad \text{(constant step)}$$

$$h_k = \frac{h}{\sqrt{k+1}}.$$

2. *Full relaxation*:

$$h_k = \arg\min_{h \geq 0} f(x_k - hf'(x_k)).$$

3. *Goldstein–Armijo* rule: Find $x_{k+1} = x_k - hf'(x_k)$ such that

$$\alpha\langle f'(x_k), x_k - x_{k+1}\rangle \leq f(x_k) - f(x_{k+1}), \tag{1.2.10}$$

$$\beta\langle f'(x_k), x_k - x_{k+1}\rangle \geq f(x_k) - f(x_{k+1}), \tag{1.2.11}$$

where $0 < \alpha < \beta < 1$ are some fixed parameters.

Comparing these strategies, we see that the first strategy is the simplest one. Indeed, it is often used, but mainly in the context of convex optimization. In that framework the behavior of functions is much more predictable than in the general nonlinear case.

The second strategy is completely theoretical. It is never used in practice since even in one-dimensional cases we cannot find an exact minimum in finite time.

The third strategy is used in the majority of the practical algorithms. It has the following geometric interpretation. Let us fix $x \in R^n$. Consider the function of one variable

$$\phi(h) = f(x - hf'(x)), \quad h \geq 0.$$

Then the step-size values acceptable for this strategy belong to the part of the graph of ϕ that is located between two linear functions:

$$\phi_1(h) = f(x) - \alpha h \parallel f'(x) \parallel^2, \quad \phi_2(h) = f(x) - \beta h \parallel f'(x) \parallel^2 .$$

Note that $\phi(0) = \phi_1(0) = \phi_2(0)$ and $\phi'(0) < \phi_2'(0) < \phi_1'(0) < 0$. Therefore, the acceptable values exist unless $\phi(h)$ is not bounded below. There are several very fast one-dimensional procedures for finding a point satisfying the conditions of this strategy, but their description is not so important for us now.

Let us estimate the performance of the gradient method. Consider the problem

$$\min_{x \in R^n} f(x),$$

with $f \in C_L^{1,1}(R^n)$. And assume that $f(x)$ is bounded below on R^n.

Let us evaluate a result of one gradient step. Consider $y = x - hf'(x)$. Then, in view of (1.2.5), we have

$$\begin{aligned} f(y) &\leq f(x) + \langle f'(x), y - x\rangle + \tfrac{L}{2} \parallel y - x \parallel^2 \\ &= f(x) - h \parallel f'(x) \parallel^2 + \tfrac{h^2}{2} L \parallel f'(x) \parallel^2 \qquad (1.2.12) \\ &= f(x) - h(1 - \tfrac{h}{2}L) \parallel f'(x) \parallel^2 . \end{aligned}$$

Thus, in order to get the best estimate for possible decrease of the objective function, we have to solve the following one-dimensional problem:

$$\Delta(h) = -h\left(1 - \tfrac{h}{2}L\right) \to \min_h .$$

Computing the derivative of this function, we conclude that the optimal step size must satisfy the equation $\Delta'(h) = hL - 1 = 0$. Thus, that is $h^* = \frac{1}{L}$, which is a minimum of $\Delta(h)$ since $\Delta''(h) = L > 0$.

Thus, our considerations prove that one step of the gradient method decreases the value of objective function at least as follows:

$$f(y) \leq f(x) - \tfrac{1}{2L} \parallel f'(x) \parallel^2 .$$

Let us check what is going on with the above step-size strategies.

Let $x_{k+1} = x_k - h_k f'(x_k)$. Then for the constant step strategy, $h_k = h$, we have

$$f(x_k) - f(x_{k+1}) \geq h(1 - \tfrac{1}{2}Lh) \parallel f'(x_k) \parallel^2 .$$

Therefore, if we choose $h_k = \frac{2\alpha}{L}$ with $\alpha \in (0,1)$, then

$$f(x_k) - f(x_{k+1}) \geq \tfrac{2}{L}\alpha(1-\alpha) \parallel f'(x_k) \parallel^2 .$$

Of course, the optimal choice is $h_k = \frac{1}{L}$.

For the full relaxation strategy we have

$$f(x_k) - f(x_{k+1}) \geq \tfrac{1}{2L} \parallel f'(x_k) \parallel^2$$

since the maximal decrease is not worse than that for $h_k = \frac{1}{L}$.

Finally, for the Goldstein–Armijo rule in view of (1.2.11) we have

$$f(x_k) - f(x_{k+1}) \leq \beta\langle f'(x_k), x_k - x_{k+1}\rangle = \beta h_k \parallel f'(x_k) \parallel^2 .$$

From (1.2.12) we obtain

$$f(x_k) - f(x_{k+1}) \geq h_k \left(1 - \tfrac{h_k}{2}L\right) \parallel f'(x_k) \parallel^2 .$$

Therefore $h_k \geq \frac{2}{L}(1-\beta)$. Further, using (1.2.10) we have

$$f(x_k) - f(x_{k+1}) \geq \alpha\langle f'(x_k), x_k - x_{k+1}\rangle = \alpha h_k \parallel f'(x_k) \parallel^2 .$$

Combining this inequality with the previous one, we conclude that

$$f(x_k) - f(x_{k+1}) \geq \tfrac{2}{L}\alpha(1-\beta) \parallel f'(x_k) \parallel^2 .$$

Thus, we have proved that in *all* cases we have

$$f(x_k) - f(x_{k+1}) \geq \tfrac{\omega}{L} \parallel f'(x_k) \parallel^2, \tag{1.2.13}$$

where ω is some positive constant.

Now we are ready to estimate the performance of the gradient scheme. Let us sum up the inequalities (1.2.13) for $k = 0 \dots N$. We obtain

$$\tfrac{\omega}{L} \sum_{k=0}^{N} \parallel f'(x_k) \parallel^2 \leq f(x_0) - f(x_{N+1}) \leq f(x_0) - f^*, \tag{1.2.14}$$

where f^* is the optimal value of the problem (1.2.1). As a simple consequence of (1.2.14) we have

$$\parallel f'(x_k) \parallel \to 0 \quad \text{as} \quad k \to \infty.$$

However, we can also say something about the *convergence rate*. Indeed, denote

$$g_N^* = \min_{0 \le k \le N} g_k,$$

where $g_k = \| f'(x_k) \|$. Then, in view of (1.2.14), we come to the following inequality:

$$g_N^* \le \frac{1}{\sqrt{N+1}} \left[\frac{1}{\omega} L(f(x_0) - f^*) \right]^{1/2}. \tag{1.2.15}$$

The right-hand side of this inequality describes the *rate of convergence* of the sequence $\{g_N^*\}$ to zero. Note that we cannot say anything about the rate of convergence of sequences $\{f(x_k)\}$ and $\{x_k\}$.

Recall, that in general nonlinear optimization our goal is quite moderate: We want to find only a local minimum of our problem. Nevertheless, even this goal is unreachable for a gradient method. Let us consider the following example.

EXAMPLE 1.2.2 Let us look at the following function of two variables:

$$f(x) \equiv f(x^{(1)}, x^{(2)}) = \tfrac{1}{2}(x^{(1)})^2 + \tfrac{1}{4}(x^{(2)})^4 - \tfrac{1}{2}(x^{(2)})^2.$$

The gradient of this function is $f'(x) = (x^{(1)}, (x^{(2)})^3 - x^{(2)})^T$. Therefore there are only three points which can pretend to be a local minimum of this function:

$$x_1^* = (0,0), \quad x_2^* = (0,-1), \quad x_3^* = (0,1).$$

Computing the Hessian of this function,

$$f''(x) = \begin{pmatrix} 1 & 0 \\ 0 & 3(x^{(2)})^2 - 1 \end{pmatrix},$$

we conclude that x_2^* and x_3^* are the isolated local minima[1], but x_1^* is only a *stationary point* of our function. Indeed, $f(x_1^*) = 0$ and $f(x_1^* + \epsilon e_2) = \frac{\epsilon^4}{4} - \frac{\epsilon^2}{2} < 0$ for ϵ small enough.

Now, let us consider the trajectory of the gradient method, which starts from $x_0 = (1,0)$. Note that the second coordinate of this point is zero. Therefore, the second coordinate of $f'(x_0)$ is also zero. Consequently, the second coordinate of x_1 is zero, etc. Thus, the entire sequence of points, generated by the gradient method will have the second coordinate equal to zero. This means that this sequence converges to x_1^*.

[1] In fact, in our example they are the global solutions.

To conclude our example, note that this situation is typical for all first–order unconstrained minimization methods. Without additional rather strict assumptions it is impossible to guarantee their global convergence to a local minimum, only to a stationary point. □

Note that inequality (1.2.15) provides us with an example of a new notion, that is the *rate of convergence* of minimization process. How can we use this notion in the complexity analysis? Rate of convergence delivers the *upper* complexity bounds for a problem class. These bounds are always justified by some numerical methods. If there exists a method, for which its upper complexity bounds are proportional to the *lower* complexity bounds of the problem class, we call this method *optimal.* Recall that in Section 1 we have already seen an example of optimal method.

Let us look at an example of upper complexity bounds.

EXAMPLE 1.2.3 Consider the following problem class:

Model:	1. Unconstrained minimization. 2. $f \in C_L^{1,1}(R^n)$. 3. $f(x)$ is bounded below.	(1.2.16)
Oracle:	First order black box.	
ϵ **– solution:**	$f(\bar{x}) \leq f(x_0)$, $\parallel f'(\bar{x}) \parallel \leq \epsilon$.	

Note, that inequality (1.2.15) can be used in order to obtain an upper bound for the number of steps (= calls of the oracle), which is necessary to find a point with a small norm of the gradient. For that, let us write down the following inequality:

$$g_N^* \leq \frac{1}{\sqrt{N+1}} \left[\frac{1}{\omega} L(f(x_0) - f^*)\right]^{1/2} \leq \epsilon.$$

Therefore, if $N + 1 \geq \frac{L}{\omega\epsilon^2}(f(x_0) - f^*)$, we necessarily have $g_N^* \leq \epsilon$.

Thus, we can use the value $\frac{L}{\omega\epsilon^2}(f(x_0) - f^*)$ as an *upper complexity bound* for our problem class. Comparing this estimate with the result of Theorem 1.1.2, we can see that it is much better; at least it does not depend on n. The lower complexity bound for the class (1.2.16) is not known. □

Let us check, what can be said about the *local* convergence of the gradient method. Consider the unconstrained minimization problem

$$\min_{x \in R^n} f(x)$$

under the following assumptions:

1. $f \in C_M^{2,2}(R^n)$.
2. There exists a local minimum of function f at which the Hessian is *positive definite.*
3. We know some bounds $0 < l \leq L < \infty$ for the Hessian at x^*:

$$lI_n \preceq f''(x^*) \preceq LI_n. \tag{1.2.17}$$

4. Our starting point x_0 is close enough to x^*.

Consider the process: $x_{k+1} = x_k - h_k f'(x_k)$. Note that $f'(x^*) = 0$. Hence,

$$\begin{aligned} f'(x_k) &= f'(x_k) - f'(x^*) = \int_0^1 f''(x^* + \tau(x_k - x^*))(x_k - x^*)d\tau \\ &= G_k(x_k - x^*), \end{aligned}$$

where $G_k = \int_0^1 f''(x^* + \tau(x_k - x^*))d\tau$. Therefore

$$x_{k+1} - x^* = x_k - x^* - h_k G_k(x_k - x^*) = (I - h_k G_k)(x_k - x^*).$$

There is a standard technique for analyzing processes of this type, which is based on *contracting mappings.* Let sequence $\{a_k\}$ be defined as follows:

$$a_0 \in R^n, \quad a_{k+1} = A_k a_k,$$

where A_k are $(n \times n)$ matrices such that $\| A_k \| \leq 1 - q$ with $q \in (0, 1)$. Then we can estimate the rate of convergence of sequence $\{a_k\}$ to zero:

$$\| a_{k+1} \| \leq (1 - q) \| a_k \| \leq (1 - q)^{k+1} \| a_0 \| \to 0.$$

In our case we need to estimate $\| I_n - h_k G_k \|$. Denote $r_k = \| x_k - x^* \|$. In view of Corollary 1.2.2, we have

$$f''(x^*) - \tau M r_k I_n \preceq f''(x^* + \tau(x_k - x^*)) \preceq f''(x^*) + \tau M r_k I_n.$$

Therefore, using assumption (1.2.17), we obtain

$$(l - \tfrac{r_k}{2}M)I_n \preceq G_k \preceq (L + \tfrac{r_k}{2}M)I_n.$$

Hence, $(1 - h_k(L + \frac{r_k}{2}M))I_n \leq I_n - h_k G_k \leq (1 - h_k(l - \frac{r_k}{2}M))I_n$ and we conclude that

$$\| I_n - h_k G_k \| \leq \max\{a_k(h_k), b_k(h_k)\}, \tag{1.2.18}$$

where $a_k(h) = 1 - h(l - \frac{r_k}{2}M)$ and $b_k(h) = h(L + \frac{r_k}{2}M) - 1$.

Note that $a_k(0) = 1$ and $b_k(0) = -1$. Therefore, if $r_k < \bar{r} \equiv \frac{2l}{M}$, then $a_k(h)$ is a strictly decreasing function of h and we can ensure

$$\| I_n - h_k G_k \| < 1$$

for small enough h_k. In this case we will have $r_{k+1} < r_k$.

As usual, many step-size strategies are available. For example, we can choose $h_k = \frac{1}{L}$. Let us consider the "optimal" strategy consisting in minimizing the right-hand side of (1.2.18):

$$\max\{a_k(h), b_k(h)\} \to \min_h.$$

Assume that $r_0 < \bar{r}$. Then, if we form the sequence $\{x_k\}$ using the optimal strategy, we can be sure that $r_{k+1} < r_k < \bar{r}$. Further, the optimal step size h_k^* can be found from the equation:

$$a_k(h) = b_k(h) \quad \Leftrightarrow \quad 1 - h(l - \tfrac{r_k}{2}M) = h(L + \tfrac{r_k}{2}M) - 1.$$

Hence

$$h_k^* = \tfrac{2}{L+l}. \tag{1.2.19}$$

(Surprisingly enough, the optimal step does not depend on M.) Under this choice we obtain

$$r_{k+1} \leq \tfrac{(L-l)r_k}{L+l} + \tfrac{Mr_k^2}{L+l}.$$

Let us estimate the rate of convergence of the process. Denote $q = \frac{2l}{L+l}$ and $a_k = \frac{M}{L+l}r_k$ $(< q)$. Then

$$a_{k+1} \leq (1-q)a_k + a_k^2 = a_k(1 + (a_k - q)) = \tfrac{a_k(1-(a_k-q)^2)}{1-(a_k-q)} \leq \tfrac{a_k}{1+q-a_k}.$$

Therefore $\frac{1}{a_{k+1}} \geq \frac{1+q}{a_k} - 1$, or

$$\tfrac{q}{a_{k+1}} - 1 \geq \tfrac{q(1+q)}{a_k} - q - 1 = (1+q)\left(\tfrac{q}{a_k} - 1\right).$$

Hence,

$$\begin{aligned}\frac{q}{a_k} - 1 &\geq (1+q)^k \left(\frac{q}{a_0} - 1\right) = (1+q)^k \left(\frac{2l}{L+l} \cdot \frac{L+l}{r_0 M} - 1\right) \\ &= (1+q)^k \left(\frac{\bar{r}}{r_0} - 1\right).\end{aligned}$$

Thus,

$$a_k \leq \frac{q r_0}{r_0 + (1+q)^k (\bar{r} - r_0)} \leq \frac{q r_0}{\bar{r} - r_0} \left(\frac{1}{1+q}\right)^k.$$

This proves the following theorem.

THEOREM 1.2.4 *Let function $f(x)$ satisfy our assumptions and let the starting point x_0 be close enough to a local minimum:*

$$r_0 = \| x_0 - x^* \| < \bar{r} = \frac{2l}{M}.$$

Then the gradient method with step size (1.2.19) converges as follows:

$$\| x_k - x^* \| \leq \frac{\bar{r} r_0}{\bar{r} - r_0} \left(1 - \frac{2l}{L+3l}\right)^k.$$

This rate of convergence is called *linear.*

1.2.4 Newton method

The Newton method is widely known as a technique for finding a root of a function of one variable. Let $\phi(t) : R \to R$. Consider the equation

$$\phi(t^*) = 0.$$

The Newton method is based on linear approximation. Assume that we get some t close enough to t^*. Note that

$$\phi(t + \Delta t) = \phi(t) + \phi'(t)\Delta t + o(| \Delta t |).$$

Therefore the equation $\phi(t + \Delta t) = 0$ can be approximated by the following *linear* equation:

$$\phi(t) + \phi'(t)\Delta t = 0.$$

We can expect that the solution of this equation, the displacement Δt, is a good approximation to the optimal displacement $\Delta t^* = t^* - t$. Converting this idea in an algorithmic form, we get the process

$$t_{k+1} = t_k - \frac{\phi(t_k)}{\phi'(t_k)}.$$

This scheme can be naturally extended onto the problem of finding solution to a system of nonlinear equations,

$$F(x) = 0,$$

where $x \in R^n$ and $F(x) : R^n \to R^n$. In this case we have to define the displacement Δx as a solution to the following system of linear equations:

$$F(x) + F'(x)\Delta x = 0$$

(it is called the *Newton system*). If the Jacobian $F'(x)$ is nondegenerate, we can compute displacement $\Delta x = -[F'(x)]^{-1}F(x)$. The corresponding iterative scheme looks as follows:

$$x_{k+1} = x_k - [F'(x_k)]^{-1}F(x_k).$$

Finally, in view of Theorem 1.2.1, we can replace the unconstrained minimization problem by a problem of finding roots of the nonlinear system

$$f'(x) = 0. \tag{1.2.20}$$

(This replacement is not completely equivalent, but it works in nondegenerate situations.) Further, for solving (1.2.20) we can apply a standard Newton method for systems of nonlinear equations. In this case, the Newton system looks as follows:

$$f'(x) + f''(x)\Delta x = 0.$$

Hence, the Newton method for optimization problems appears to be in the form

$$\boxed{x_{k+1} = x_k - [f''(x_k)]^{-1}f'(x_k).} \tag{1.2.21}$$

Note that we can obtain the process (1.2.21), using the idea of quadratic approximation. Consider this approximation, computed with respect to the point x_k:

$$\phi(x) = f(x_k) + \langle f'(x_k), x - x_k\rangle + \tfrac{1}{2}\langle f''(x_k)(x - x_k), x - x_k\rangle.$$

Assume that $f''(x_k) \succ 0$. Then we can choose x_{k+1} as a point of minimum of the quadratic function $\phi(x)$. This means that

$$\phi'(x_{k+1}) = f'(x_k) + f''(x_k)(x_{k+1} - x_k) = 0,$$

and we come again to the Newton process (1.2.21).

We will see that the convergence of the Newton method in a neighborhood of a strict local minimum is very fast. However, this method has two serious drawbacks. Firstly, it can break down if $f''(x_k)$ is degenerate. Secondly, the Newton process can diverge. Let us look at the following example.

EXAMPLE 1.2.4 Let us apply the Newton method for finding a root of the following function of one variable:

$$\phi(t) = \frac{t}{\sqrt{1+t^2}}.$$

Clearly, $t^* = 0$. Note that

$$\phi'(t) = \frac{1}{[1+t^2]^{3/2}}.$$

Therefore the Newton process looks as follows:

$$t_{k+1} = t_k - \frac{\phi(t_k)}{\phi'(t_k)} = t_k - \frac{t_k}{\sqrt{1+t_k^2}} \cdot [1 + t_k^2]^{3/2} = -t_k^3.$$

Thus, if $| \ t_0 \ | < 1$, then this method converges and the convergence is extremely fast. The points ± 1 are the oscillation points of this method. If $| \ t_0 \ | > 1$, then the method diverges. □

In order to avoid a possible divergence, in practice we can apply a *damped Newton method*:

$$\boxed{x_{k+1} = x_k - h_k[f''(x_k)]^{-1} f'(x_k),}$$

where $h_k > 0$ is a step-size parameter. At the initial stage of the method we can use the same step size strategies as for the gradient scheme. At the final stage it is reasonable to choose $h_k = 1$.

Let us study the local convergence of the Newton method. Consider the problem

$$\min_{x \in R^n} f(x)$$

under the following assumptions:

1. $f \in C_M^{2,2}(R^n)$.

2. There exists a local minimum of function f with *positive definite* Hessian:

$$f''(x^*) \succeq lI_n, \quad l > 0. \qquad (1.2.22)$$

3. Our starting point x_0 is close enough to x^*.

Consider the process: $x_{k+1} = x_k - [f''(x_k)]^{-1} f'(x_k)$. Then, using the same reasoning as for the gradient method, we obtain the following representation:

$$\begin{aligned} x_{k+1} - x^* &= x_k - x^* - [f''(x_k)]^{-1} f'(x_k) \\ &= x_k - x^* - [f''(x_k)]^{-1} \int_0^1 f''(x^* + \tau(x_k - x^*))(x_k - x^*) d\tau \\ &= [f''(x_k)]^{-1} G_k (x_k - x^*), \end{aligned}$$

where $G_k = \int_0^1 [f''(x_k) - f''(x^* + \tau(x_k - x^*))] d\tau$.

Denote $r_k = \| x_k - x^* \|$. Then

$$\begin{aligned} \| G_k \| &= \| \int_0^1 [f''(x_k) - f''(x^* + \tau(x_k - x^*))] d\tau \| \\ &\leq \int_0^1 \| f''(x_k) - f''(x^* + \tau(x_k - x^*)) \| d\tau \\ &\leq \int_0^1 M(1 - \tau) r_k d\tau = \frac{r_k}{2} M. \end{aligned}$$

In view of Corollary 1.2.2, and (1.2.22), we have

$$f''(x_k) \geq f''(x^*) - M r_k I_n \geq (l - M r_k) I_n.$$

Therefore, if $r_k < \frac{l}{M}$, then $f''(x_k)$ is positive definite and

$$\| [f''(x_k)]^{-1} \| \leq (l - M r_k)^{-1}.$$

Hence, for r_k small enough ($r_k < \frac{2l}{3M}$), we have

$$r_{k+1} \leq \frac{M r_k^2}{2(l - M r_k)} \quad (< r_k).$$

The rate of convergence of this type is called *quadratic*.

Thus, we have proved the following theorem.

THEOREM 1.2.5 *Let function $f(x)$ satisfy our assumptions. Suppose that the initial starting point x_0 is close enough to x^*:*

$$\| x_0 - x^* \| < \bar{r} = \frac{2l}{3M}.$$

Then $\| x_k - x^* \| < \bar{r}$ for all k and the Newton method converges quadratically:

$$\| x_{k+1} - x^* \| \leq \frac{M \| x_k - x^* \|^2}{2(l - M \| x_k - x^* \|)}.$$

Comparing this result with the rate of convergence of the gradient method, we see that the Newton method is much faster. Surprisingly enough, the *region of quadratic convergence* of the Newton method is almost the same as the region of the linear convergence of the gradient method. This justifies a standard recommendation to use the gradient method only at the initial stage of the minimization process in order to get close to a local minimum. The final job should be performed by the Newton method.

In this section we have seen several examples of the convergence rate. Let us make a correspondence between these rates and the complexity bounds. As we have seen in Example 1.2.3, the upper bound for the analytical complexity of a problem class is an inverse function of the rate of convergence.

1. *Sublinear rate.* This rate is described in terms of a power function of the iteration counter. For example, we can have $r_k \leq \frac{c}{\sqrt{k}}$. In this case the upper complexity bound of corresponding problem class justified by this scheme is $\left(\frac{c}{\epsilon}\right)^2$.

   Sublinear rate is rather slow. In terms of complexity, each new right digit of the answer takes the amount of computations *comparable* with the total amount of the previous work. Note also, that the constant c plays a significant role in the corresponding complexity estimate.

2. *Linear rate.* This rate is given in terms of an exponential function of the iteration counter. For example,

$$r_k \leq c(1-q)^k.$$

   Note that the corresponding complexity bound is $\frac{1}{q}(\ln c + \ln \frac{1}{\epsilon})$.

   This rate is fast: Each new right digit of the answer takes a constant amount of computations. Moreover, the dependence of the complexity estimate in constant c is very weak.

3. *Quadratic rate.* This rate has a form of a double exponential function of the iteration counter. For example,

$$r_{k+1} \leq c r_k^2.$$

   The corresponding complexity estimate depends on a double logarithm of the desired accuracy: $\ln \ln \frac{1}{\epsilon}$.

   This rate is extremely fast: Each iteration doubles the number of right digits in the answer. The constant c is important only for the starting moment of the quadratic convergence ($cr_k < 1$).

1.3 First-order methods in nonlinear optimization

(Gradient method and Newton method: What is different? Idea of variable metric; Variable metric methods; Conjugate gradient methods; Constrained minimization: Penalty functions and penalty function methods; Barrier functions and barrier function methods.)

1.3.1 Gradient method and Newton method: What is different?

In the previous section we have considered two local methods for finding a local minimum in the simplest minimization problem

$$\min_{x \in R^n} f(x),$$

with $f \in C_L^{2,2}(R^n)$. Those are the gradient method

$$x_{k+1} = x_k - h_k f'(x_k), \quad h_k > 0.$$

and the Newton Method:

$$x_{k+1} = x_k - [f''(x_k)]^{-1} f'(x_k).$$

Recall that the local rate of convergence of these methods is different. We have seen, that the gradient method has a linear rate and the Newton method converges quadratically. What is the reason for this difference?

If we look at the analytic form of these methods, we can see at least the following formal difference: In the gradient method the search direction is the antigradient, while in the Newton method we multiply the antigradient by some matrix, that is the inverse Hessian. Let us try to derive these directions using some "universal" reasoning.

Let us fix some $\bar{x} \in R^n$. Consider the following approximation of the function $f(x)$:

$$\phi_1(x) = f(\bar{x}) + \langle f'(\bar{x}), x - \bar{x}\rangle + \frac{1}{2h} \| x - \bar{x} \|^2,$$

where the parameter h is positive. The first-order optimality condition provides us with the following equation for x_1^*, the unconstrained minimum of the function $\phi_1(x)$:

$$\phi_1'(x_1^*) = f'(\bar{x}) + \frac{1}{h}(x_1^* - \bar{x}) = 0.$$

Thus, $x_1^* = \bar{x} - hf'(\bar{x})$. That is exactly the iterate of the gradient method. Note, that if $h \in (0, \frac{1}{L}]$, then the function $\phi_1(x)$ is a *global upper* approximation of $f(x)$:

$$f(x) \le \phi_1(x), \quad \forall x \in R^n,$$

(see Lemma 1.2.3). This fact is responsible for global convergence of the gradient method.

Further, consider a quadratic approximation of function $f(x)$:

$$\phi_2(x) = f(\bar{x}) + \langle f'(\bar{x}), x - \bar{x}\rangle + \tfrac{1}{2}\langle f''(\bar{x})(x - \bar{x}), x - \bar{x}\rangle.$$

We have already seen that the minimum of this function is

$$x_2^* = \bar{x} - [f''(\bar{x})]^{-1} f'(\bar{x}),$$

and that is exactly the iterate of the Newton method.

Thus, we can try to use some approximations of function $f(x)$, which are better than $\phi_1(x)$ and which are less expensive than $\phi_2(x)$.

Let G be a positive definite $n \times n$-matrix. Denote

$$\phi_G(x) = f(\bar{x}) + \langle f'(\bar{x}), x - \bar{x}\rangle + \tfrac{1}{2}\langle G(x - \bar{x}), x - \bar{x}\rangle.$$

Computing its minimum from the equation

$$\phi'_G(x_G^*) = f'(\bar{x}) + G(x_G^* - \bar{x}) = 0,$$

we obtain

$$x_G^* = \bar{x} - G^{-1} f'(\bar{x}). \qquad (1.3.1)$$

The first-order methods, which form a sequence of matrices

$$\{G_k\} : G_k \to f''(x^*)$$

(or $\{H_k\} : H_k \equiv G_k^{-1} \to [f''(x^*)]^{-1}$), are called the *variable metric* methods. (Sometimes the name *quasi-Newton* methods is used.) In these methods only the gradients are involved in the process of generating the sequences $\{G_k\}$ or $\{H_k\}$.

The updating rule (1.3.1) is very common in optimization. Let us provide it with one more interpretation.

Note that the gradient and the Hessian of a nonlinear function $f(x)$ are defined *with respect to* a standard Euclidean inner product on R^n:

$$\langle x, y\rangle = \sum_{i=1}^{n} x^{(i)} y^{(i)}, \quad x, y \in R^n, \qquad \| x \| = \langle x, x\rangle^{1/2}.$$

Indeed, the definition of the gradient is

$$f(x + h) = f(x) + \langle f'(x), h\rangle + o(\| h \|),$$

and from this equation we derive its coordinate representation:

$$f'(x) = \left(\frac{\partial f(x)}{\partial x^{(1)}}, \ldots, \frac{\partial f(x)}{\partial x^{(n)}}\right)^T.$$

Let us introduce now a new inner product. Consider a symmetric positive definite $n \times n$-matrix A. For $x,\ y \in R^n$ denote

$$\langle x, y\rangle_A = \langle Ax, y\rangle, \qquad \| x \|_A = \langle Ax, x\rangle^{1/2}.$$

The function $\| x \|_A$ is a new norm on R^n. Note that topologically this new metric is equivalent to the old one:

$$\lambda_n(A)^{1/2} \| x \| \le \| x \|_A \le \lambda_1(A)^{1/2} \| x \|,$$

where $\lambda_n(A)$ and $\lambda_1(A)$ are the smallest and the largest eigenvalues of the matrix A. However, the gradient and the Hessian, computed with respect to the new inner product are changing:

$$\begin{aligned} f(x+h) &= f(x) + \langle f'(x), h\rangle + \tfrac{1}{2}\langle f''(x)h, h\rangle + o(\| h \|) \\ &= f(x) + \langle A^{-1}f'(x), h\rangle_A + \tfrac{1}{2}\langle A^{-1}f''(x)h, h\rangle_A + o(\| h \|_A). \end{aligned}$$

Hence, $f'_A(x) = A^{-1}f'(x)$ is the new gradient and $f''_A(x) = A^{-1}f''(x)$ is the new Hessian.

Thus, the direction used in the Newton method can be seen as a gradient computed with respect to the metric defined by $A = f''(x)$. Note that the Hessian of $f(x)$ at x computed with respect to $A = f''(x)$ is I_n.

EXAMPLE 1.3.1 Consider quadratic function

$$f(x) = \alpha + \langle a, x\rangle + \tfrac{1}{2}\langle Ax, x\rangle,$$

where $A = A^T \succ 0$. Note that $f'(x) = Ax + a$, $f''(x) = A$ and

$$f'(x^*) = Ax^* + a = 0$$

for $x^* = -A^{-1}a$. Let us compute the Newton direction at some $x \in R^n$:

$$d_N(x) = [f''(x)]^{-1}f'(x) = A^{-1}(Ax + a) = x + A^{-1}a.$$

Therefore for any $x \in R^n$ we have $x - d_N(x) = -A^{-1}a = x^*$. Thus, for a quadratic function the Newton method converges in one step. Note also that

$$\begin{aligned} f(x) &= \alpha + \langle A^{-1}a, x\rangle_A + \tfrac{1}{2} \| x \|_A^2, \\ f'_A(x) &= A^{-1}f'(x) = d_N(x), \\ f''_A(x) &= A^{-1}f''(x) = I_n. \end{aligned}$$

□

Let us write down a general scheme of the *variable metric* methods.

Variable metric method
0. Choose $x_0 \in R^n$. Set $H_0 = I_n$. Compute $f(x_0)$ and $f'(x_0)$.
1. kth iteration ($k \geq 0$). a). Set $p_k = H_k f'(x_k)$. b). Find $x_{k+1} = x_k - h_k p_k$ (see Section 1.2.3 for step-size rules). c). Compute $f(x_{k+1})$ and $f'(x_{k+1})$. d). Update the matrix H_k : $H_k \to H_{k+1}$.

The variable metric schemes differ one from another only in implementation of Step 1d), which updates matrix H_k. For that, they use new information, accumulated at Step 1c), namely the gradient $f'(x_{k+1})$. The idea is justified by the following property of a quadratic function. Let

$$f(x) = \alpha + \langle a, x \rangle + \tfrac{1}{2}\langle Ax, x \rangle, \quad f'(x) = Ax + a.$$

Then, for any $x, y \in R^n$ we have $f'(x) - f'(y) = A(x - y)$. This identity explains the origin of the so-called *quasi-Newton rule.*

Quasi-Newton rule
Choose H_{k+1} such that $H_{k+1}(f'(x_{k+1}) - f'(x_k)) = x_{k+1} - x_k.$

Actually, there are many ways to satisfy this relation. Below we present several examples of the schemes that usually are recommended as the most efficient ones.

EXAMPLE 1.3.2 Denote

$$\Delta H_k = H_{k+1} - H_k, \quad \gamma_k = f'(x_{k+1}) - f'(x_k), \quad \delta_k = x_{k+1} - x_k.$$

Then the quasi-Newton relation is satisfied by the following rules.

1. *Rank-one correction* scheme.

$$\Delta H_k = \frac{(\delta_k - H_k\gamma_k)(\delta_k - H_k\gamma_k)^T}{\langle \delta_k - H_k\gamma_k, \gamma_k \rangle}.$$

2. *Davidon–Fletcher–Powell* scheme (DFP).

$$\Delta H_k = \frac{\delta_k \delta_k^T}{\langle \gamma_k, \delta_k \rangle} - \frac{H_k \gamma_k \gamma_k^T H_k}{\langle H_k \gamma_k, \gamma_k \rangle}.$$

3. *Broyden–Fletcher–Goldfarb–Shanno* scheme (BFGS).

$$\Delta H_k = \frac{H_k \gamma_k \delta_k^T + \delta_k \gamma_k^T H_k}{\langle H_k \gamma_k, \gamma_k \rangle} - \beta_k \frac{H_k \gamma_k \gamma_k^T H_k}{\langle H_k \gamma_k, \gamma_k \rangle},$$

where $\beta_k = 1 + \langle \gamma_k, \delta_k \rangle / \langle H_k \gamma_k, \gamma_k \rangle$.

Clearly, there are many other possibilities. From the computational point of view, BFGS is considered as the most stable scheme. □

Note that for quadratic functions the variable metric methods usually terminate in n iterations. In a neighborhood of strict minimum they have a *superlinear* rate of convergence: for any $x_0 \in R^n$ there exists a number N such that for all $k \geq N$ we have

$$\| x_{k+1} - x^* \| \leq \text{const} \cdot \| x_k - x^* \| \cdot \| x_{k-n} - x^* \|$$

(the proofs are very long and technical). As far as global convergence is concerned, these methods are not better than the gradient method (at least, from the theoretical point of view).

Note that in the variable metric schemes it is necessary to store and update a symmetric $n \times n$-matrix. Thus, each iteration needs $O(n^2)$ auxiliary arithmetic operations. During many years this feature was considered as one of the main drawbacks of the variable metric methods. That stimulated the interest in so-called *conjugate gradients* schemes, which have much lower complexity of each iteration (see Section 1.3.2). However, in view of an amazing growth of computer power in the last decades, these objections are not so important anymore.

1.3.2 Conjugate gradients

The conjugate gradients methods were initially proposed for minimizing a quadratic function. Consider the problem

$$\min_{x \in R^n} f(x), \tag{1.3.2}$$

with $f(x) = \alpha + \langle a, x \rangle + \frac{1}{2}\langle Ax, x \rangle$ and $A = A^T \succ 0$. We have already seen that the solution of this problem is $x^* = -A^{-1}a$. Therefore, our objective function can be written in the following form:

$$\begin{aligned} f(x) &= \alpha + \langle a, x \rangle + \tfrac{1}{2}\langle Ax, x \rangle = \alpha - \langle Ax^*, x \rangle + \tfrac{1}{2}\langle Ax, x \rangle \\ &= \alpha - \tfrac{1}{2}\langle Ax^*, x^* \rangle + \tfrac{1}{2}\langle A(x - x^*), x - x^* \rangle. \end{aligned}$$

Thus, $f^* = \alpha - \frac{1}{2}\langle Ax^*, x^* \rangle$ and $f'(x) = A(x - x^*)$.

Suppose we are given by a starting point x_0. Consider the linear *Krylov* subspaces

$$\mathcal{L}_k = \text{Lin}\{A(x_0 - x^*), \dots, A^k(x_0 - x^*)\}, \quad k \geq 1,$$

where A^k is the kth power of matrix A. The sequence of points $\{x_k\}$ generated by a *conjugate gradients method* is defined as follows:

$$\boxed{x_k = \arg\min\{f(x) \mid x \in x_0 + \mathcal{L}_k\}, \; k \geq 1.} \tag{1.3.3}$$

This definition looks quite artificial. However, later we will see that this method can be written in a pure "algorithmic" form. We need representation (1.3.3) only for theoretical analysis.

LEMMA 1.3.1 *For any $k \geq 1$ we have $\mathcal{L}_k = \text{Lin}\{f'(x_0), \dots, f'(x_{k-1})\}$.*

Proof: For $k = 1$ the statement is true since $f'(x_0) = A(x_0 - x^*)$. Suppose that it is true for some $k \geq 1$. Then

$$x_k = x_0 + \sum_{i=1}^{k} \lambda^{(i)} A^i (x_0 - x^*)$$

with some $\lambda \in R^k$. Therefore

$$f'(x_k) = A(x_0 - x^*) + \sum_{i=1}^{k} \lambda^{(i)} A^{i+1}(x_0 - x^*) = y + \lambda^{(k)} A^{k+1}(x_0 - x^*),$$

for certain y from $\mathcal{L}_k$. Thus,

$$\begin{aligned} \mathcal{L}_{k+1} &\equiv \text{Lin}\{\mathcal{L}_k, A^{k+1}(x_0 - x^*)\} = \text{Lin}\{\mathcal{L}_k, f'(x_k)\} \\ &= \text{Lin}\{f'(x_0), \dots, f'(x_k)\}. \end{aligned}$$

□

The next result helps to understand the behavior of the sequence $\{x_k\}$.

LEMMA 1.3.2 *For any k, $i \geq 0$, $k \neq i$ we have $\langle f'(x_k), f'(x_i)\rangle = 0$.*

Proof: Let $k > i$. Consider the function

$$\phi(\lambda) = f\left(x_0 + \sum_{j=1}^{k} \lambda^{(j)} f'(x_{j-1})\right), \quad \lambda \in R^k.$$

In view of Lemma 1.3.1, for some λ_* we have $x_k = x_0 + \sum_{j=1}^{k} \lambda_*^{(j)} f'(x_{j-1})$. However, by definition, x_k is the point of minimum of $f(x)$ on $\mathcal{L}_k$. Therefore $\phi'(\lambda_*) = 0$. It remains to compute the components of the gradient:

$$0 = \frac{\partial \phi(\lambda_*)}{\partial \lambda^{(i)}} = \langle f'(x_k), f'(x_i)\rangle.$$

□

COROLLARY 1.3.1 *The sequence generated by the conjugate gradients method for (1.3.2) is finite.*

Proof: The number of orthogonal directions in R^n cannot exceed n. □

COROLLARY 1.3.2 *For any $p \in \mathcal{L}_k$ we have $\langle f'(x_k), p\rangle = 0$.* □

The last auxiliary result explains the name of the method. Denote $\delta_i = x_{i+1} - x_i$. It is clear that $\mathcal{L}_k = \text{Lin}\{\delta_0, \ldots, \delta_{k-1}\}$.

LEMMA 1.3.3 *For any $k \neq i$ we have $\langle A\delta_k, \delta_i\rangle = 0$.*

(Such directions are called *conjugate* with respect to A.)
Proof: Without loss of generality we can assume that $k > i$. Then

$$\langle A\delta_k, \delta_i\rangle = \langle A(x_{k+1} - x_k), \delta_i\rangle = \langle f'(x_{k+1}) - f'(x_k), \delta_i\rangle = 0$$

since $\delta_i = x_{i+1} - x_i \in \mathcal{L}_{i+1} \subseteq \mathcal{L}_k$. □

Let us show how we can write down the conjugate gradients method in a more algorithmic form. Since $\mathcal{L}_k = \text{Lin}\{\delta_0, \ldots, \delta_{k-1}\}$, we can represent x_{k+1} as follows:

$$x_{k+1} = x_k - h_k f'(x_k) + \sum_{j=0}^{k-1} \lambda^{(j)} \delta_j.$$

In our notation that is

$$\delta_k = -h_k f'(x_k) + \sum_{j=0}^{k-1} \lambda^{(j)} \delta_j. \tag{1.3.4}$$

Let us compute the coefficients of the representation. Multiplying (1.3.4) by A and δ_i, $0 \leq i \leq k-1$, and using Lemma 1.3.3 we obtain

$$\begin{aligned} 0 &= \langle A\delta_k, \delta_i \rangle = -h_k \langle Af'(x_k), \delta_i \rangle + \sum_{j=0}^{k-1} \lambda^{(j)} \langle A\delta_j, \delta_i \rangle \\ &= -h_k \langle Af'(x_k), \delta_i \rangle + \lambda^{(i)} \langle A\delta_i, \delta_i \rangle \\ &= -h_k \langle f'(x_k), f'(x_{i+1}) - f'(x_i) \rangle + \lambda^{(i)} \langle A\delta_i, \delta_i \rangle. \end{aligned}$$

Hence, in view of Lemma 1.3.2, $\lambda_i = 0$, $i < k-1$. For $i = k-1$ we have

$$\lambda^{(k-1)} = \frac{h_k \|f'(x_k)\|^2}{\langle A\delta_{k-1}, \delta_{k-1} \rangle} = \frac{h_k \|f'(x_k)\|^2}{\langle f'(x_k) - f'(x_{k-1}), \delta_{k-1} \rangle}.$$

Thus, $x_{k+1} = x_k - h_k p_k$, where

$$p_k = f'(x_k) - \frac{\|f'(x_k)\|^2 \delta_{k-1}}{\langle f'(x_k) - f'(x_{k-1}), \delta_{k-1} \rangle} = f'(x_k) - \frac{\|f'(x_k)\|^2 p_{k-1}}{\langle f'(x_k) - f'(x_{k-1}), p_{k-1} \rangle}$$

since $\delta_{k-1} = -h_{k-1} p_{k-1}$ by the definition of $\{p_k\}$.

Note that we managed to write down a conjugate gradients scheme in terms of the gradients of the objective function $f(x)$. This provides us with a possibility to apply *formally* this scheme for minimizing a general nonlinear function. Of course, such extension destroys all properties of the process, which are specific for the quadratic functions. However, in the neighborhood of a strict local minimum the objective function is close to quadratic. Therefore asymptotically this method can be fast.

Let us present a general scheme of the conjugate gradients method for minimizing a nonlinear function.

Conjugate gradient method
0. Let $x_0 \in R^n$. Compute $f(x_0), f'(x_0)$. Set $p_0 = f'(x_0)$.
1. kth iteration ($k \geq 0$). a). Find $x_{k+1} = x_k + h_k p_k$ (by "exact" line search). b). Compute $f(x_{k+1})$ and $f'(x_{k+1})$. c). Compute the coefficient β_k. d). Set $p_{k+1} = f'(x_{k+1}) - \beta_k p_k$.

In that scheme we did not specify yet the coefficient β_k. In fact, there are many different formulas for this coefficient. All of them give the same result on quadratic functions, but in a general nonlinear case they generate different sequences. Let us present three of the most popular expressions.

1. $\beta_k = \frac{\| f'(x_{k+1}) \|^2}{\langle f'(x_{k+1}) - f'(x_k), p_k \rangle}$.

2. *Fletcher–Rieves:* $\beta_k = -\frac{\| f'(x_{k+1}) \|^2}{\| f'(x_k) \|^2}$.

3. *Polak–Ribbiere:* $\beta_k = -\frac{\langle f'(x_{k+1}), f'(x_{k+1}) - f'(x_k) \rangle}{\| f'(x_k) \|^2}$.

Recall that in the quadratic case the conjugate gradients method terminates in n iterations (or less). Algorithmically, this means that $p_{n+1} = 0$. In a nonlinear case that is not true. However, after n iteration this direction loses any interpretation. Therefore, in all practical schemes there exists a *restarting* strategy, which at some moment sets $\beta_k = 0$ (usually after every n iterations). This ensures a global convergence of the scheme (since we have a usual gradient step just after the restart and all other iterations decrease the function value). In a neighborhood of a strict minimum the conjugate gradients schemes have a local n-step quadratic convergence:

$$\| x_{n+1} - x^* \| \leq \text{const} \cdot \| x_0 - x^* \|^2 .$$

Note, that this local convergence is slower than that of the variable metric methods. However, the conjugate gradients schemes have an advantage of a very cheap iteration. As far as the global convergence is concerned, the conjugate gradients, in general, are not better than the gradient method.

1.3.3 Constrained minimization

Let us discuss briefly the main ideas underlying the methods of general constrained minimization. The problem we deal with is as follows:

$$\begin{aligned} f_0(x) &\rightarrow \min, \\ f_i(x) &\leq 0, \; i = 1 \ldots m. \end{aligned} \tag{1.3.5}$$

where $f_i(x)$ are smooth functions. For example, we can consider $f_i(x)$ from $C_L^{1,1}(R^n)$.

Since the components of the problem (1.3.5) are general nonlinear functions, we cannot expect that this problem is easier than an unconstrained minimization problem. Indeed, even the standard difficulties with stationary points, which we have in unconstrained minimization, appear in (1.3.5) in a much stronger form. Note that a stationary point of this problem (whatever it is) can be infeasible for the system of functional constraints. Hence, any minimization scheme attracted by such a point should accept that it fails even to find a feasible solution to (1.3.5).

Therefore, the following reasoning looks quite convincing.

1. We have efficient methods for unconstrained minimization. (?)[2]
2. Unconstrained minimization is simpler than the constrained one. (?)[3]
3. Therefore, let us try to approximate a solution to the problem (1.3.5) by a sequence of solutions to some auxiliary unconstrained minimization problems.

This philosophy is implemented by the schemes of *Sequential Unconstrained Minimization*. There are two main groups of such methods: the *penalty function* methods and the *barrier* methods. Let us describe the basic ideas of these approaches.

[2] In fact, that is not absolutely true. We will see, that in order to apply an unconstrained minimization method for solving constrained problems, we need to be able to find at least a strict local minimum. And we have already seen (Example 1.2.2), that this could pose a problem.

[3] We are not going to discuss the correctness of this statement for general nonlinear problems. We just prevent the reader from extending it onto another problem classes. In the next chapters we will have a possibility to see that this statement is not always true.

We start from penalty function methods.

DEFINITION 1.3.1 *A continuous function* $\Phi(x)$ *is called a* penalty function *for a closed set* Q *if*

- $\Phi(x) = 0$ *for any* $x \in Q$,
- $\Phi(x) > 0$ *for any* $x \notin Q$.

Sometimes a penalty function is called just *penalty*. The main property of the penalty functions is as follows.

> If $\Phi_1(x)$ is a penalty for Q_1 and $\Phi_2(x)$ is a penalty for Q_2, then $\Phi_1(x) + \Phi_2(x)$ is a penalty for intersection $Q_1 \bigcap Q_2$.

Let us give several examples of such functions.

EXAMPLE 1.3.3 Denote $(a)_+ = \max\{a, 0\}$. Let

$$Q = \{x \in R^n \mid f_i(x) \leq 0,\ i = 1 \dots m\}.$$

Then the following functions are penalties for Q:

1. *Quadratic penalty:* $\Phi(x) = \sum\limits_{i=1}^{m} (f_i(x))_+^2$.

2. *Nonsmooth penalty:* $\Phi(x) = \sum\limits_{i=1}^{m} (f_i(x))_+$.

The reader can easily continue the list. □

The general scheme of a penalty function method is as follows.

Penalty function method

0. Choose $x_0 \in R^n$. Choose a sequence of penalty coefficients: $0 < t_k < t_{k+1}$ and $t_k \to \infty$.

1. kth iteration ($k \geq 0$).
Find a point $x_{k+1} = \arg \min\limits_{x \in R^n} \{f_0(x) + t_k \Phi(x)\}$
using x_k as a starting point.

It is easy to prove the convergence of this scheme assuming that x_{k+1} is a global minimum of the auxiliary function.[4] Denote

$$\Psi_k(x) = f_0(x) + t_k\Phi(x), \quad \Psi_k^* = \min_{x \in R^n} \Psi_k(x).$$

(Ψ_k^* is the *global* optimal value of $\Psi_k(x)$). Denote by x^* the global solution to (1.3.5).

THEOREM 1.3.1 *Let there exist a value $\bar{t} > 0$ such that the set*

$$S = \{x \in R^n \mid f_0(x) + \bar{t}\Phi(x) \leq f_0(x^*)\}$$

is bounded. Then

$$\lim_{k \to \infty} f(x_k) = f_0(x^*), \quad \lim_{k \to \infty} \Phi(x_k) = 0.$$

Proof: Note that $\Psi_k^* \leq \Psi_k(x^*) = f_0(x^*)$. At the same time, for any $x \in R^n$ we have $\Psi_{k+1}(x) \geq \Psi_k(x)$. Therefore $\Psi_{k+1}^* \geq \Psi_k^*$. Thus, there exists a limit $\lim\limits_{k \to \infty} \Psi_k^* \equiv \Psi^* \leq f^*$. If $t_k > \bar{t}$ then

$$f_0(x_k) + \bar{t}\Phi(x_k) \leq f_0(x_k) + t_k\Phi(x_k) = \Psi_k^* \leq f_0(x^*).$$

Therefore, the sequence $\{x_k\}$ has limit points. Since $\lim\limits_{k \to \infty} t_k = +\infty$, for any such point x_* we have $\Phi(x_*) = 0$ and $f_0(x_*) \leq f_0(x^*)$. Thus $x_* \in Q$ and

$$\Psi^* = f_0(x_*) + \Phi(x_*) = f_0(x_*) \geq f_0(x^*).$$

□

Note that this result is very general, but not too informative. There are still many questions, which should be answered. For example, we do not know what kind of penalty function we should use. What should be the rules for choosing the penalty coefficients? What should be the accuracy for solving the auxiliary problems? The main feature of these questions is that they can be hardly addressed in the framework of general nonlinear optimization theory. Traditionally, they are considered as questions to be answered by computational practice.

Let us look at the barrier methods.

DEFINITION 1.3.2 *Let Q be a closed set with nonempty interior. A continuous function $F(x)$ is called a* barrier function *for Q if $F(x) \to \infty$ when x approaches the boundary of Q.*

---

[4] If we assume that it is a strict local minimum, then the result is much weaker.

Sometimes a barrier function is called *barrier* for short. Similarly to the penalty functions, the barriers possess the following property:

If $F_1(x)$ is a barrier for Q_1 and $F_2(x)$ is a barrier for Q_2, then $F_1(x) + F_2(x)$ is a barrier for intersection $Q_1 \bigcap Q_2$.

In order to apply the barrier approach, the problem (1.3.5) must satisfy the *Slater condition:*

$$\exists \bar{x} : \quad f_i(\bar{x}) < 0, \quad i = 1 \dots m.$$

Let us look at some examples of barrier functions.

EXAMPLE 1.3.4 Let $Q = \{x \in R^n \mid f_i(x) \leq 0, \ i = 1 \dots m\}$. Then all functions below are barriers for Q:

1. *Power-function* barrier: $F(x) = \sum\limits_{i=1}^{m} \frac{1}{(-f_i(x))^p}$, $p \geq 1$.

2. *Logarithmic* barrier: $F(x) = -\sum\limits_{i=1}^{m} \ln(-f_i(x))$.

3. *Exponential* barrier: $F(x) = \sum\limits_{i=1}^{m} \exp\left(\frac{1}{-f_i(x)}\right)$.

The reader can easily extend this list. □

The scheme of a barrier method is as follows.

Barrier function method
0. Choose $x_0 \in \operatorname{int} Q$. Choose a sequence of penalty coefficients: $0 < t_k < t_{k+1}$ and $t_k \to \infty$.
1. kth iteration ($k \geq 0$). Find a point $x_{k+1} = \arg\min\limits_{x \in Q} \{f_0(x) + \frac{1}{t_k} F(x)\}$ using x_k as a starting point.

Let us prove the convergence of this method assuming that x_{k+1} is a global minimum of the auxiliary function. Denote

$$\Psi_k(x) = f_0(x) + \tfrac{1}{t_k} F(x), \quad \Psi_k^* = \min_{x \in Q} \Psi_k(x),$$

(Ψ_k^* is the global optimal value of $\Psi_k(x)$). And let f^* be the optimal value of the problem (1.3.5).

THEOREM 1.3.2 *Let barrier $F(x)$ be bounded below on Q. Then*

$$\lim_{k\to\infty} \Psi_k^* = f^*.$$

Proof: Let $F(x) \geq F^*$ for all $x \in Q$. For arbitrary $\bar{x} \in \operatorname{int} Q$ we have

$$\sup \lim_{k\to\infty} \Psi_k^* \leq \lim_{k\to\infty} \left[f_0(\bar{x}) + \tfrac{1}{t_k} F(\bar{x}) \right] = f_0(\bar{x}).$$

Therefore $\sup \lim_{k\to\infty} \Psi_k^* \leq f^*$. On the other hand,

$$\Psi_k^* = \min_{x\in Q} \left\{ f_0(x) + \tfrac{1}{t_k} F(x) \right\} \geq \min_{x\in Q} \left\{ f_0(x) + \tfrac{1}{t_k} F^* \right\} = f^* + \tfrac{1}{t_k} F^*.$$

Thus, $\lim_{k\to\infty} \Psi_k^* = f^*$. □

The same as with the penalty functions method, there are many questions to be answered. We do not know how to find the starting point x_0 and how to choose the best barrier function. We do not know the rules for updating the penalty coefficients and the acceptable accuracy of the solutions to the auxiliary problems. Finally, we have no idea about the efficiency estimates of this process. And the reason is not in the lack of the theory. Our problem (1.3.5) is just too complicated. We will see that all of the above questions get *precise* answers in the framework of convex optimization.

We have finished our brief presentation of general nonlinear optimization. It was really very short and there are many interesting theoretical topics that we did not mention. That is because the main goal of this book is to describe the areas of optimization in which we can obtain some clear and complete results on the performance of numerical methods. Unfortunately, the general nonlinear optimization is just too complicated to fit the goal. However, it is impossible to skip this field since a lot of basic ideas, underlying the convex optimization methods, have their origin in general nonlinear optimization theory. The gradient method and the Newton method, sequential unconstrained minimization and barrier functions were originally developed and used for general optimization problems. But only the framework of convex optimization allows these ideas to get their real power. In the next chapters of this book we will see many examples of the second birth of these old ideas.

Chapter 2

SMOOTH CONVEX OPTIMIZATION

2.1 Minimization of smooth functions

(Smooth convex functions; Lower complexity bounds for $\mathcal{F}_L^{\infty,1}(R^n)$; Strongly convex functions. Lower complexity bounds $\mathcal{S}_{\mu,L}^{\infty,1}(R^n)$; Gradient method.)

2.1.1 Smooth convex functions

In this section we deal with unconstrained minimization problem

$$\min_{x \in R^n} f(x), \tag{2.1.1}$$

where the function $f(x)$ is smooth enough. Recall that in the previous chapter we were trying to solve this problem under very weak assumptions on function f. And we have seen that in this general situation we cannot do too much: It is impossible to guarantee convergence even to a local minimum, impossible to get acceptable bounds on the global performance of minimization schemes, etc. Let us try to introduce some reasonable assumptions on function f to make our problem more tractable. For that, let us try to determine the desired properties of a class of differentiable functions $\mathcal{F}$ we want to work with.

From the results of the previous chapter we can get an impression that the main reasons of our troubles is the weakness of the first-order optimality condition (Theorem 1.2.1). Indeed, we have seen that, in general, the gradient method converges only to a stationary point of function f (see inequality (1.2.15) and Example 1.2.2). Therefore the first additional property we definitely need is as follows.

ASSUMPTION 2.1.1 *For any $f \in \mathcal{F}$ the first-order optimality condition is sufficient for a point to be a global solution to (2.1.1).*

Further, the main feature of any tractable functional class $\mathcal{F}$ is the possibility to verify inclusion $f \in \mathcal{F}$ in a simple way. Usually that is ensured by a set of *basic elements* of the class and by the list of possible *operations* with elements of $\mathcal{F}$, which keep the result in the class (such operations are called *invariant*). An excellent example is the class of differentiable functions: In order to check either a function is differentiable or not, we need just to look at its analytical expression.

We do not want to restrict our class too much. Therefore, let us introduce only one invariant operation for the hypothetical class $\mathcal{F}$.

ASSUMPTION 2.1.2 *If $f_1, f_2 \in \mathcal{F}$ and $\alpha, \beta \geq 0$, then $\alpha f_1 + \beta f_2 \in \mathcal{F}$.*

The reason for the restriction on the sign of coefficients in this assumption is evident: We would like to see x^2 in our class, but function $-x^2$ is not suitable for our goals.

Finally, let us add in $\mathcal{F}$ some basic elements.

ASSUMPTION 2.1.3 *Any linear function $f(x) = \alpha + \langle a, x \rangle$ belongs to $\mathcal{F}$.*[1]

Note that the linear function $f(x)$ perfectly fits Assumption 2.1.1. Indeed, $f'(x) = 0$ implies that this function is constant and any point in R^n is its global minimum.

It turns out that we have assumed enough to specify our functional class. Consider $f \in \mathcal{F}$. Let us fix some $x_0 \in R^n$ and consider the function

$$\phi(y) = f(y) - \langle f'(x_0), y \rangle.$$

Then $\phi \in \mathcal{F}$ in view of Assumptions 2.1.2 and 2.1.3. Note that

$$\phi'(y) \mid_{y=x_0} = f'(x_0) - f'(x_0) = 0.$$

Therefore, in view of Assumption 2.1.1, x_0 is the global minimum of function ϕ and for any $y \in R^n$ we have

$$\phi(y) \geq \phi(x_0) = f(x_0) - \langle f'(x_0), x_0 \rangle.$$

Hence, $f(y) \geq f(x_0) + \langle f'(x_0), x - x_0 \rangle$.

This inequality is very well known in optimization. It defines the class of differentiable *convex* functions.

DEFINITION 2.1.1 *A continuously differentiable function $f(x)$ is called* convex *on R^n (notation $f \in \mathcal{F}^1(R^n)$) if for any $x, y \in R^n$ we have*

$$f(y) \geq f(x) + \langle f'(x), y - x \rangle. \qquad (2.1.2)$$

[1] This is not a description of the whole set of basic elements. We just say that we want to have linear functions in our class.

If $-f(x)$ is convex, we call $f(x)$ *concave.*

In what follows we consider also the classes of convex functions $\mathcal{F}_L^{k,l}(Q)$ with the same meaning of the indices as for the classes $C_L^{k,l}(Q)$.

Let us check our assumptions, which become now the *properties* of the functional class.

THEOREM 2.1.1 *If $f \in \mathcal{F}^1(R^n)$ and $f'(x^*) = 0$ then x^* is the global minimum of $f(x)$ on R^n.*

Proof: In view of inequality (2.1.2), for any $x \in R^n$ we have

$$f(x) \geq f(x^*) + \langle f'(x^*), x - x^* \rangle = f(x^*).$$

□

Thus, we get what we want in Assumption 2.1.1. Let us check Assumption 2.1.2.

LEMMA 2.1.1 *If f_1 and f_2 belong to $\mathcal{F}^1(R^n)$ and $\alpha, \beta \geq 0$ then function $f = \alpha f_1 + \beta f_2$ also belongs to $\mathcal{F}^1(R^n)$.*

Proof: For any $x, y \in R^n$ we have

$$\begin{aligned} f_1(y) &\geq f_1(x) + \langle f_1'(x), y - x \rangle, \\ f_2(y) &\geq f_2(x) + \langle f_2'(x), y - x \rangle. \end{aligned}$$

It remains to multiply the first equation by α, the second one by β and add the results. □

Thus, for differentiable functions our hypothetical class coincides with the class of convex functions. Let us present their main properties.

The next statement significantly increases our possibilities in *constructing* the convex functions.

LEMMA 2.1.2 *If $f \in \mathcal{F}^1(R^m)$, $b \in R^m$ and $A : R^n \to R^m$ then*

$$\phi(x) = f(Ax + b) \in \mathcal{F}^1(R^n).$$

Proof: Indeed, let $x, y \in R^n$. Denote $\bar{x} = Ax + b$, $\bar{y} = Ay + b$. Since $\phi'(x) = A^T f'(Ax + b)$, we have

$$\begin{aligned}
\phi(y) = f(\bar{y}) &\geq f(\bar{x}) + \langle f'(\bar{x}), \bar{y} - \bar{x} \rangle \\
&= \phi(x) + \langle f'(\bar{x}), A(y - x) \rangle \\
&= \phi(x) + \langle A^T f'(\bar{x}), y - x \rangle \\
&= \phi(x) + \langle \phi'(x), y - x \rangle.
\end{aligned}$$

□

In order to simplify the verification of inclusion $f \in \mathcal{F}^1(R^n)$, we provide this class with several equivalent definitions.

THEOREM 2.1.2 *Continuously differentiable function f belongs to the class $\mathcal{F}^1(R^n)$ if and only if for any $x, y \in R^n$ and $\alpha \in [0,1]$ we have*[2]

$$f(\alpha x + (1-\alpha)y) \leq \alpha f(x) + (1-\alpha) f(y). \qquad (2.1.3)$$

Proof: Denote $x_\alpha = \alpha x + (1-\alpha)y$. Let $f \in \mathcal{F}^1(R^n)$. Then

$$\begin{aligned}
f(x_\alpha) &\leq f(y) + \langle f'(x_\alpha), y - x_\alpha \rangle &&= f(y) + \alpha \langle f'(x_\alpha), y - x \rangle, \\
f(x_\alpha) &\leq f(x) + \langle f'(x_\alpha), x - x_\alpha \rangle &&= f(x) - (1-\alpha) \langle f'(x_\alpha), y - x \rangle.
\end{aligned}$$

Multiplying first inequality by $(1-\alpha)$, the second one by α and adding the results, we get (2.1.3).

Let (2.1.3) be true for all $x, y \in R^n$ and $\alpha \in [0,1]$. Let us choose some $\alpha \in [0,1)$. Then

$$\begin{aligned}
f(y) &\geq \tfrac{1}{1-\alpha}[f(x_\alpha) - \alpha f(x)] = f(x) + \tfrac{1}{1-\alpha}[f(x_\alpha) - f(x)] \\
&= f(x) + \tfrac{1}{1-\alpha}[f(x + (1-\alpha)(y - x)) - f(x)].
\end{aligned}$$

Tending α to 1, we get (2.1.2). □

THEOREM 2.1.3 *Continuously differentiable function f belongs to the class $\mathcal{F}^1(R^n)$ if and only if for any $x, y \in R^n$ we have*

$$\langle f'(x) - f'(y), x - y \rangle \geq 0. \qquad (2.1.4)$$

[2] Note that inequality (2.1.3) without assumption on differentiability of f, serves as a definition of *general convex* functions. We will study these functions in detail in the next chapter.

Proof: Let f be a convex continuously differentiable function. Then

$$f(x) \geq f(y) + \langle f'(y), x - y\rangle, \quad f(y) \geq f(x) + \langle f'(x), y - x\rangle.$$

Adding these inequalities, we get (2.1.4).

Let (2.1.4) hold for all $x, y \in R^n$. Denote $x_\tau = x + \tau(y - x)$. Then

$$\begin{aligned} f(y) &= f(x) + \int_0^1 \langle f'(x + \tau(y-x)), y - x\rangle d\tau \\ &= f(x) + \langle f'(x), y - x\rangle + \int_0^1 \langle f'(x_\tau) - f'(x), y - x\rangle d\tau \\ &= f(x) + \langle f'(x), y - x\rangle + \int_0^1 \tfrac{1}{\tau}\langle f'(x_\tau) - f'(x), x_\tau - x\rangle d\tau \\ &\geq f(x) + \langle f'(x), y - x\rangle. \end{aligned}$$

□

Sometimes it is more convenient to work with functions from the class $\mathcal{F}^2(R^n) \subset \mathcal{F}^1(R^n)$.

THEOREM 2.1.4 *Two times continuously differentiable function f belongs to $\mathcal{F}^2(R^n)$ if and only for any $x \in R^n$ we have*

$$f''(x) \succeq 0. \tag{2.1.5}$$

Proof: Let $f \in C^2(R^n)$ be convex. Denote $x_\tau = x + \tau s, \quad \tau > 0$. Then, in view of (2.1.4), we have

$$\begin{aligned} 0 &\leq \tfrac{1}{\tau}\langle f'(x_\tau) - f'(x), x_\tau - x\rangle = \tfrac{1}{\tau}\langle f'(x_\tau) - f'(x), s\rangle \\ &= \tfrac{1}{\tau}\int_0^\tau \langle f''(x + \lambda s)s, s\rangle d\lambda, \end{aligned}$$

and we get (2.1.5) by tending τ to zero.

Let (2.1.5) hold for all $x \in R^n$. Then

$$\begin{aligned} f(y) &= f(x) + \langle f'(x), y - x\rangle \\ &\quad + \int_0^1\int_0^\tau \langle f''(x + \lambda(y-x))(y-x), y - x\rangle d\lambda d\tau \\ &\geq f(x) + \langle f'(x), y - x\rangle. \end{aligned}$$

□

Let us look at some examples of differentiable convex functions.

EXAMPLE 2.1.1 1. Linear function $f(x) = \alpha + \langle a, x \rangle$ is convex.

2. Let a matrix A be symmetric and positive semidefinite. Then the quadratic function

$$f(x) = \alpha + \langle a, x \rangle + \tfrac{1}{2}\langle Ax, x \rangle$$

is convex (since $f''(x) = A \succeq 0$).

3. The following functions of one variable belong to $\mathcal{F}^1(R)$:

$$\begin{aligned} f(x) &= e^x, \\ f(x) &= \mid x \mid^p, \quad p > 1, \\ f(x) &= \tfrac{x^2}{1-|x|}, \\ f(x) &= \mid x \mid - \ln(1+ \mid x \mid). \end{aligned}$$

We can check that using Theorem 2.1.4.

Therefore, the function arising in *geometric optimization*,

$$f(x) = \sum_{i=1}^{m} e^{\alpha_i + \langle a_i, x \rangle},$$

is convex (see Lemma 2.1.2). Similarly, the function arising in l_p*-norm* approximation problem,

$$f(x) = \sum_{i=1}^{m} \mid \langle a_i, x \rangle - b_i \mid^p,$$

is convex too. □

As with general nonlinear functions, the differentiability itself cannot ensure any special topological properties of convex functions. Therefore we need to consider the problem classes with Lipschitz continuous derivatives of a certain order. The most important class of that type is $\mathcal{F}_L^{1,1}(R^n)$, the class of convex functions with Lipschitz continuous gradient. Let us provide it with several necessary and sufficient conditions.

THEOREM 2.1.5 *All conditions below, holding for all x, $y \in R^n$ and α from $[0,1]$, are equivalent to inclusion $f \in \mathcal{F}_L^{1,1}(R^n)$:*

$$0 \le f(y) - f(x) - \langle f'(x), y - x \rangle \le \tfrac{L}{2} \parallel x - y \parallel^2, \qquad (2.1.6)$$

$$f(x)+\langle f'(x),y-x\rangle+\tfrac{1}{2L}\parallel f'(x)-f'(y)\parallel^2\le f(y), \qquad (2.1.7)$$

$$\tfrac{1}{L}\parallel f'(x)-f'(y)\parallel^2\le\langle f'(x)-f'(y),x-y\rangle, \qquad (2.1.8)$$

$$\langle f'(x)-f'(y),x-y\rangle\le L\parallel x-y\parallel^2, \qquad (2.1.9)$$

$$\begin{aligned}\alpha f(x)+(1-\alpha)f(y) &\ge f(\alpha x+(1-\alpha)y)\\ &\quad+\tfrac{\alpha(1-\alpha)}{2L}\parallel f'(x)-f'(y)\parallel^2,\end{aligned} \qquad (2.1.10)$$

$$\begin{aligned}\alpha f(x)+(1-\alpha)f(y) &\le f(\alpha x+(1-\alpha)y)\\ &\quad+\alpha(1-\alpha)\tfrac{L}{2}\parallel x-y\parallel^2 .\end{aligned} \qquad (2.1.11)$$

Proof: Indeed, (2.1.6) follows from the definition of convex functions and Lemma 1.2.3. Further, let us fix $x_0\in R^n$. Consider the function

$$\phi(y)=f(y)-\langle f'(x_0),y\rangle.$$

Note that $\phi\in\mathcal{F}_L^{1,1}(R^n)$ and its optimal point is $y^*=x_0$. Therefore, in view of (2.1.6), we have

$$\phi(y^*)\le\phi(y-\tfrac{1}{L}\phi'(y))\le\phi(y)-\tfrac{1}{2L}\parallel\phi'(y)\parallel^2 .$$

And we get (2.1.7) since $\phi'(y)=f'(y)-f'(x_0)$.

We obtain (2.1.8) from inequality (2.1.7) by adding two copies of it with x and y interchanged. Applying now to (2.1.8) Cauchy–Schwartz inequality we get $\parallel f'(x)-f'(y)\parallel\le L\parallel x-y\parallel$.

In the same way we can obtain (2.1.9) from (2.1.6). In order to get (2.1.6) from (2.1.9) we apply integration:

$$\begin{aligned}f(y)-f(x)-\langle f'(x),y-x\rangle &=\int_0^1\langle f'(x+\tau(y-x))-f'(x),y-x\rangle d\tau\\ &\le\tfrac{1}{2}L\|y-x\|^2.\end{aligned}$$

Let us prove the last two inequalities. Denote $x_\alpha=\alpha x+(1-\alpha)y$. Then, using (2.1.7) we get

$$f(x)\ \ge\ f(x_\alpha)+\langle f'(x_\alpha),(1-\alpha)(x-y)\rangle+\tfrac{1}{2L}\parallel f'(x)-f'(x_\alpha)\parallel^2,$$

$$f(y)\ \ge\ f(x_\alpha)+\langle f'(x_\alpha),\alpha(y-x)\rangle+\tfrac{1}{2L}\parallel f'(y)-f'(x_\alpha)\parallel^2 .$$

Adding these inequalities multiplied by α and $(1-\alpha)$ respectively, and using inequality

$$\alpha \parallel g_1 - u \parallel^2 + (1-\alpha) \parallel g_2 - u \parallel^2 \geq \alpha(1-\alpha) \parallel g_1 - g_2 \parallel^2,$$

we get (2.1.10). It is easy to check that we get (2.1.7) from (2.1.10) by tending $\alpha \to 1$.

Similarly, from (2.1.6) we get

$$\begin{aligned} f(x) &\leq f(x_\alpha) + \langle f'(x_\alpha), (1-\alpha)(x-y)\rangle + \tfrac{L}{2} \parallel (1-\alpha)(x-y) \parallel^2, \\ f(y) &\leq f(x_\alpha) + \langle f'(x_\alpha), \alpha(y-x)\rangle + \tfrac{L}{2} \parallel \alpha(y-x) \parallel^2. \end{aligned}$$

Adding these inequalities multiplied by α and $(1-\alpha)$ respectively, we obtain (2.1.11). And we get back to (2.1.6) as $\alpha \to 1$. □

Finally, let us give a characterization of the class $\mathcal{F}_L^{2,1}(R^n)$.

THEOREM 2.1.6 *Two times continuously differentiable function f belongs to $\mathcal{F}_L^{2,1}(R^n)$ if and only for any $x \in R^n$ we have*

$$0 \preceq f''(x) \preceq LI_n. \qquad (2.1.12)$$

Proof: The statement follows from Theorem 2.1.4 and (2.1.9). □

2.1.2 Lower complexity bounds for $\mathcal{F}_L^{\infty,1}(R^n)$

Before we go forward with optimization methods, let us check our possibilities in minimizing smooth convex functions. In this section we obtain the lower complexity bounds for optimization problems with objective functions from $\mathcal{F}_L^{\infty,1}(R^n)$ (and, consequently, $\mathcal{F}_L^{1,1}(R^n)$).

Recall that our problem class is as follows.

Model:	$\min\limits_{x \in R^n} f(x), \quad f \in \mathcal{F}_L^{1,1}(R^n).$
Oracle:	First-order local black box.
Approximate solution:	$\bar{x} \in R^n, \ f(\bar{x}) - f^* \leq \epsilon.$

In order to make our considerations simpler, let us introduce the following assumption on iterative processes.

ASSUMPTION 2.1.4 *An iterative method $\mathcal{M}$ generates a sequence of test points $\{x_k\}$ such that*

$$x_k \in x_0 + \mathrm{Lin}\,\{f'(x_0), \ldots, f'(x_{k-1})\}, \quad k \geq 1.$$

This assumption is not absolutely necessary and it can be avoided by a more sophisticated reasoning. However, it holds for the majority of practical methods.

We can prove the lower complexity bounds for our problem class without developing a resisting oracle. Instead, we just point out the "worst function in the world" (that means, in $\mathcal{F}_L^{\infty,1}(R^n)$). This function appears to be difficult for *all* iterative schemes satisfying Assumption 2.1.4.

Let us fix some constant $L > 0$. Consider the following family of quadratic functions

$$f_k(x) = \frac{L}{4}\left\{\frac{1}{2}[(x^{(1)})^2 + \sum_{i=1}^{k-1}(x^{(i)} - x^{(i+1)})^2 + (x^{(k)})^2] - x^{(1)}\right\}$$

for $k = 1 \ldots n$. Note that for all $s \in R^n$ we have

$$\langle f_k''(x)s, s\rangle = \frac{L}{4}\left[(s^{(1)})^2 + \sum_{i=1}^{k-1}(s^{(i)} - s^{(i+1)})^2 + (s^{(k)})^2\right] \geq 0,$$

and

$$\begin{aligned}\langle f_k''(x)s, s\rangle &\leq \frac{L}{4}[(s^{(1)})^2 + \sum_{i=1}^{k-1} 2((s^{(i)})^2 + (s^{(i+1)})^2) + (s^{(k)})^2] \\ &\leq L\sum_{i=1}^{n}(s^{(i)})^2.\end{aligned}$$

Thus, $0 \preceq f_k''(x) \preceq LI_n$. Therefore $f_k(x) \in \mathcal{F}_L^{\infty,1}(R^n)$, $1 \leq k \leq n$.

Let us compute the minimum of function f_k. It is easy to see that $f_k''(x) = \frac{L}{4}A_k$ with

$$A_k = \left(\begin{array}{c|c} k \text{ lines}\left\{\begin{array}{ccccccc} 2 & -1 & 0 & & & & \\ -1 & 2 & -1 & & 0 & & \\ 0 & -1 & 2 & & & & \\ & \ldots & & & \ldots & & \\ & 0 & & & -1 & 2 & -1 \\ & & & & 0 & -1 & 2 \end{array}\right. & 0_{n-k,k} \\ \hline 0_{n-k,k} & 0_{n-k,n-k} \end{array}\right),$$

where $0_{k,p}$ is a $(k \times p)$ zero matrix. Therefore the equation

$$f_k'(x) = A_k x - e_1 = 0$$

has the following unique solution:

$$\bar{x}_k^{(i)} = \begin{cases} 1 - \frac{i}{k+1}, & i = 1 \ldots k, \\ 0, & k+1 \leq i \leq n. \end{cases}$$

Hence, the optimal value of function f_k is

$$\begin{aligned} f_k^* &= \tfrac{L}{4}\left[\tfrac{1}{2}\langle A_k \bar{x}_k, \bar{x}_k\rangle - \langle e_1, \bar{x}_k\rangle\right] = -\tfrac{L}{8}\langle e_1, \bar{x}_k\rangle \\ &= \tfrac{L}{8}\left(-1 + \tfrac{1}{k+1}\right). \end{aligned} \tag{2.1.13}$$

Note also that

$$\sum_{i=1}^{k} i^2 = \frac{k(k+1)(2k+1)}{6} \leq \frac{(k+1)^3}{3}. \tag{2.1.14}$$

Therefore

$$\begin{aligned} \| \bar{x}_k \|^2 &= \sum_{i=1}^{n} \left(\bar{x}_k^{(i)}\right)^2 = \sum_{i=1}^{k} \left(1 - \tfrac{i}{k+1}\right)^2 \\ &= k - \tfrac{2}{k+1} \sum_{i=1}^{k} i + \tfrac{1}{(k+1)^2} \sum_{i=1}^{k} i^2 \\ &\leq k - \tfrac{2}{k+1} \cdot \tfrac{k(k+1)}{2} + \tfrac{1}{(k+1)^2} \cdot \tfrac{(k+1)^3}{3} = \tfrac{1}{3}(k+1). \end{aligned} \tag{2.1.15}$$

Denote $R^{k,n} = \{x \in R^n \mid x^{(i)} = 0,\ k+1 \leq i \leq n\}$; that is a subspace of R^n, in which only the first k components of the point can differ from zero. From the analytical form of the functions $\{f_k\}$ it is easy to see that for all $x \in R^{k,n}$ we have

$$f_p(x) = f_k(x), \quad p = k \ldots n.$$

Let us fix some p, $1 \leq p \leq n$.

LEMMA 2.1.3 *Let $x_0 = 0$. Then for any sequence $\{x_k\}_{k=0}^{p}$ satisfying the condition*

$$x_k \in \mathcal{L}_k = \text{Lin}\{f_p'(x_0), \ldots, f_p'(x_{k-1})\},$$

we have $\mathcal{L}_k \subseteq R^{k,n}$.

Proof: Since $x_0 = 0$, we have $f_p'(x_0) = -\frac{L}{4}e_1 \in R^{1,n}$. Thus $\mathcal{L}_1 \equiv R^{1,n}$.

Let $\mathcal{L}_k \subseteq R^{k,n}$ for some $k < p$. Since A_p is three-diagonal, for any $x \in R^{k,n}$ we have $f_p'(x) \in R^{k+1,n}$. Therefore $\mathcal{L}_{k+1} \subseteq R^{k+1,n}$, and we can complete the proof by induction. □

COROLLARY 2.1.1 *For any sequence* $\{x_k\}_{k=0}^p$ *such that* $x_0 = 0$ *and* $x_k \in \mathcal{L}_k$ *we have*

$$f_p(x_k) \geq f_k^*.$$

Proof: Indeed, $x_k \in \mathcal{L}_k \subseteq R^{k,n}$ and therefore $f_p(x_k) = f_k(x_k) \geq f_k^*$. □

Now we are ready to prove the main result of this section.

THEOREM 2.1.7 *For any* k, $1 \leq k \leq \frac{1}{2}(n-1)$, *and any* $x_0 \in R^n$ *there exists a function* $f \in \mathcal{F}_L^{\infty,1}(R^n)$ *such that for any first-order method* $\mathcal{M}$ *satisfying Assumption 2.1.4 we have*

$$f(x_k) - f^* \geq \frac{3L\|x_0 - x^*\|^2}{32(k+1)^2},$$

$$\| x_k - x^* \|^2 \geq \frac{1}{8} \| x_0 - x^* \|^2,$$

where x^* *is the minimum of* $f(x)$ *and* $f^* = f(x^*)$.

Proof: It is clear that the methods of this type are invariant with respect to a simultaneous shift of all objects in the space of variables. Thus, the sequence of iterates, which is generated by such a method for function $f(x)$ starting from x_0, is just a shift of the sequence generated for $\bar{f}(x) = f(x + x_0)$ starting from the origin. Therefore, we can assume that $x_0 = 0$.

Let us prove the first inequality. For that, let us fix k and apply $\mathcal{M}$ to minimizing $f(x) = f_{2k+1}(x)$. Then $x^* = \bar{x}_{2k+1}$ and $f^* = f_{2k+1}^*$. Using Corollary 2.1.1, we conclude that

$$f(x_k) \equiv f_{2k+1}(x_k) = f_k(x_k) \geq f_k^*.$$

Hence, since $x_0 = 0$, in view of (2.1.13) and (2.1.15) we get the following estimate:

$$\frac{f(x_k) - f^*}{\|x_0 - x^*\|^2} \geq \frac{\frac{L}{8}\left(-1 + \frac{1}{k+1} + 1 - \frac{1}{2k+2}\right)}{\frac{1}{3}(2k+2)} = \frac{3}{8}L \cdot \frac{1}{4(k+1)^2}.$$

Let us prove the second inequality. Since $x_k \in R^{k,n}$ and $x_0 = 0$, we have

$$\begin{aligned}\| x_k - x^* \|^2 &\geq \sum_{i=k+1}^{2k+1} \left(\bar{x}^{(i)}_{2k+1}\right)^2 = \sum_{i=k+1}^{2k+1} \left(1 - \frac{i}{2k+2}\right)^2 \\ &= k + 1 - \frac{1}{k+1} \sum_{i=k+1}^{2k+1} i + \frac{1}{4(k+1)^2} \sum_{i=k+1}^{2k+1} i^2.\end{aligned}$$

In view of (2.1.14), we have

$$\begin{aligned}\sum_{i=k+1}^{2k+1} i^2 &= \frac{1}{6}[(2k+1)(2k+2)(4k+3) - k(k+1)(2k+1)] \\ &= \frac{1}{6}(k+1)(2k+1)(7k+6).\end{aligned}$$

Therefore, using (2.1.15) we finally obtain

$$\begin{aligned}\| x_k - x^* \|^2 &\geq k + 1 - \frac{1}{k+1} \cdot \frac{(3k+2)(k+1)}{2} + \frac{(2k+1)(7k+6)}{24(k+1)} \\ &= \frac{(2k+1)(7k+6)}{24(k+1)} - \frac{k}{2} = \frac{2k^2+7k+6}{24(k+1)} \\ &\geq \frac{2k^2+7k+6}{16(k+1)^2} \| x_0 - \bar{x}_{2k+1} \|^2 \geq \frac{1}{8} \| x_0 - x^* \|^2 .\end{aligned}$$

□

The above theorem is valid only under assumption that the number of steps of the iterative scheme is not too large as compared with the dimension of the space ($k \leq \frac{1}{2}(n-1)$). The complexity bounds of that type are called *uniform* in the dimension of variables. Clearly, they are valid for very large problems, in which we cannot wait even for n iterates of the method. However, even for problems with a moderate dimension, these bounds also provide us with some information. Firstly, they describe the potential performance of numerical methods on the initial stage of the minimization process. And secondly, they warn us that without a direct use of finite-dimensional arguments we cannot get better complexity for any numerical scheme.

To conclude this section, let us note that the obtained lower bound for the value of the objective function is rather optimistic. Indeed, after one hundred iterations we could decrease the initial residual in 10^4 times. However, the result on the behavior of the minimizing sequence is quite disappointing: The convergence to the optimal point can be *arbitrarily* slow. Since that is a lower bound, this conclusion is inevitable for our problem class. The only thing we can do is to try to find problem classes in which the situation could be better. That is the goal of the next section.

2.1.3 Strongly convex functions

Thus, we are looking for a restriction of the functional class $\mathcal{F}_L^{1,1}(R^n)$, for which we can guarantee a reasonable rate of convergence to a unique solution of the minimization problem

$$\min_{x \in R^n} f(x), \quad f \in \mathcal{F}^1(R^n).$$

Recall, that in Section 1.2.3 we have proved that in a small neighborhood of a nondegenerate local minimum the gradient method converges linearly. Let us try to make this non-degeneracy assumption global. Namely, let us assume that there exists some constant $\mu > 0$ such that for any $\bar{x}$ with $f'(\bar{x}) = 0$ and any $x \in R^n$ we have

$$f(x) \geq f(\bar{x}) + \tfrac{1}{2}\mu \parallel x - \bar{x} \parallel^2 .$$

Using the same reasoning as in Section 2.1.1, we obtain the class of *strongly* convex functions.

DEFINITION 2.1.2 *A continuously differentiable function $f(x)$ is called* strongly convex *on R^n (notation $f \in \mathcal{S}_\mu^1(R^n)$) if there exists a constant $\mu > 0$ such that for any x, $y \in R^n$ we have*

$$f(y) \geq f(x) + \langle f'(x), y - x \rangle + \tfrac{1}{2}\mu \parallel y - x \parallel^2 . \qquad (2.1.16)$$

Constant μ is called the convexity parameter *of function f.*

We will also consider the classes $\mathcal{S}_{\mu,L}^{k,l}(Q)$ with the same meaning of the indices k, l and L as for the class $C_L^{k,l}(Q)$.

Let us fix some properties of strongly convex functions.

THEOREM 2.1.8 *If $f \in \mathcal{S}_\mu^1(R^n)$ and $f'(x^*) = 0$, then*

$$f(x) \geq f(x^*) + \tfrac{1}{2}\mu \parallel x - x^* \parallel^2$$

for all $x \in R^n$.

Proof: Since $f'(x^*) = 0$, in view of inequality (2.1.16), for any $x \in R^n$ we have

$$\begin{aligned} f(x) &\geq f(x^*) + \langle f'(x^*), x - x^* \rangle + \tfrac{1}{2}\mu \parallel x - x^* \parallel^2 \\ &= f(x^*) + \tfrac{1}{2}\mu \parallel x - x^* \parallel^2 . \end{aligned}$$

□

The following result justifies the addition of strongly convex functions.

LEMMA 2.1.4 *If* $f_1 \in \mathcal{S}^1_{\mu_1}(R^n)$, $f_2 \in \mathcal{S}^1_{\mu_2}(R^n)$ *and* $\alpha, \beta \geq 0$, *then*

$$f = \alpha f_1 + \beta f_2 \in \mathcal{S}^1_{\alpha\mu_1+\beta\mu_2}(R^n).$$

Proof: For any $x, y \in R^n$ we have

$$\begin{aligned} f_1(y) &\geq f_1(x) + \langle f_1'(x), y - x\rangle + \tfrac{1}{2}\mu_1 \parallel y - x \parallel^2, \\ f_2(y) &\geq f_2(x) + \langle f_2'(x), y - x\rangle + \tfrac{1}{2}\mu_2 \parallel y - x \parallel^2 . \end{aligned}$$

It remains to add these equations multiplied respectively by α and β. □

Note that the class $\mathcal{S}^1_0(R^n)$ coincides with $\mathcal{F}^1(R^n)$. Therefore addition of a convex function to a strongly convex function gives a strongly convex function with the same convexity parameter.

Let us give several equivalent definitions of strongly convex functions.

THEOREM 2.1.9 *Let f be continuously differentiable. Both conditions below, holding for all x, $y \in R^n$ and $\alpha \in [0,1]$, are equivalent to inclusion $f \in \mathcal{S}^1_\mu(R^n)$:*

$$\langle f'(x) - f'(y), x - y\rangle \geq \mu \parallel x - y \parallel^2, \qquad (2.1.17)$$

$$\begin{aligned} \alpha f(x) + (1-\alpha) f(y) &\geq f(\alpha x + (1-\alpha)y) \\ &\quad + \alpha(1-\alpha)\tfrac{\mu}{2} \parallel x - y \parallel^2 . \end{aligned} \qquad (2.1.18)$$

The proof of this theorem is very similar to the proof of Theorem 2.1.5 and we leave it as an exercise for the reader.

The next statement sometimes is useful.

THEOREM 2.1.10 *If $f \in \mathcal{S}^1_\mu(R^n)$, then for any x and y from R^n we have*

$$f(y) \leq f(x) + \langle f'(x), y - x\rangle + \tfrac{1}{2\mu} \parallel f'(x) - f'(y) \parallel^2, \qquad (2.1.19)$$

$$\langle f'(x) - f'(y), x - y\rangle \leq \tfrac{1}{\mu} \parallel f'(x) - f'(y) \parallel^2 . \qquad (2.1.20)$$

Proof: Let us fix some $x \in R^n$. Consider the function

$$\phi(y) = f(y) - \langle f'(x), y\rangle \in \mathcal{S}^1_\mu(R^n).$$

Since $\phi'(x) = 0$, in view of (2.1.16) for any $y \in R^n$ we have that

$$\begin{aligned} \phi(x) &= \min_v \phi(v) \geq \min_v [\phi(y) + \langle \phi'(y), v - y \rangle + \tfrac{1}{2}\mu \|v - y\|^2] \\ &= \phi(y) - \tfrac{1}{2\mu} \|\phi'(y)\|^2, \end{aligned}$$

and that is exactly (2.1.19). Adding two copies of (2.1.19) with x and y interchanged we get (2.1.20). □

Finally, the second-order characterization of the class $\mathcal{S}^1_\mu(R^n)$ is as follows.

THEOREM 2.1.11 *Two times continuously differentiable function f belongs to the class $\mathcal{S}^2_\mu(R^n)$ if and only if $x \in R^n$*

$$f''(x) \succeq \mu I_n. \tag{2.1.21}$$

Proof: Apply (2.1.17). □

Note we can look at examples of strongly convex functions.

EXAMPLE 2.1.2 1. $f(x) = \frac{1}{2} \parallel x \parallel^2$ belongs to $\mathcal{S}^2_1(R^n)$ since $f''(x) = I_n$.

2. Let symmetric matrix A satisfy the condition: $\mu I_n \preceq A \preceq L I_n$. Then

$$f(x) = \alpha + \langle a, x \rangle + \tfrac{1}{2} \langle Ax, x \rangle \in \mathcal{S}^{\infty,1}_{\mu,L}(R^n) \subset \mathcal{S}^{1,1}_{\mu,L}(R^n)$$

since $f''(x) = A$.

Other examples can be obtained as a sum of convex and strongly convex functions. □

For us the most interesting functional class is $\mathcal{S}^{1,1}_{\mu,L}(R^n)$. This class is described by the following inequalities:

$$\langle f'(x) - f'(y), x - y \rangle \geq \mu \parallel x - y \parallel^2, \tag{2.1.22}$$

$$\parallel f'(x) - f'(y) \parallel \leq L \parallel x - y \parallel . \tag{2.1.23}$$

The value $Q_f = L/\mu \geq 1$ is called the *condition number* of function f.

It is important that the inequality (2.1.22) can be strengthened using the additional information (2.1.23).

THEOREM 2.1.12 *If $f \in \mathcal{S}^{1,1}_{\mu,L}(R^n)$, then for any x, $y \in R^n$ we have*

$$\langle f'(x) - f'(y), x - y\rangle \geq \tfrac{\mu L}{\mu+L} \parallel x - y \parallel^2 + \tfrac{1}{\mu+L} \parallel f'(x) - f'(y) \parallel^2 . \tag{2.1.24}$$

Proof: Denote $\phi(x) = f(x) - \frac{1}{2}\mu\|x\|^2$. Then $\phi'(x) = f'(x) - \mu x$; hence, by (2.1.22) and (2.1.9) $\phi \in \mathcal{F}^{1,1}_{L-\mu}(R^n)$. If $\mu = L$, then (2.1.24) is proved. If $\mu < L$, then by (2.1.8) we have

$$\langle \phi'(x) - \phi'(y), y - x\rangle \geq \tfrac{1}{L-\mu}\|\phi'(x) - \phi'(y)\|^2,$$

and that is exactly (2.1.24). □

2.1.4 Lower complexity bounds for $\mathcal{S}^{\infty,1}_{\mu,L}(R^n)$

Let us get the lower complexity bounds for unconstrained minimization of functions from the class $\mathcal{S}^{\infty,1}_{\mu,L}(R^n) \subset \mathcal{S}^{1,1}_{\mu,L}(R^n)$. Consider the following problem class.

Model:	$\min\limits_{x \in R^n} f(x), \quad f \in \mathcal{S}^{\infty,1}_{\mu,L}(R^n), \quad \mu > 0.$
Oracle:	First-order local black box.
Approximate solution:	$\bar{x}: \ f(\bar{x}) - f^* \leq \epsilon, \ \parallel \bar{x} - x^* \parallel^2 \leq \epsilon.$

As in the previous section, we consider the methods satisfying Assumption 2.1.4. We are going to find the lower complexity bounds for our problem in terms of *condition number* $Q_f = \frac{L}{\mu}$.

Note that in the description of our problem class we do not say anything about the dimension of the space of variables. Therefore formally, this class includes also the infinite-dimensional problems.

We are going to give an example of some bad function defined in the infinite-dimensional space. We could do that also in a finite dimension, but the corresponding reasoning is more complicated.

Consider $R^\infty \equiv l_2$, the space of all sequences $x = \{x^{(i)}\}_{i=1}^\infty$ with finite norm

$$\parallel x \parallel^2 = \sum_{i=1}^{\infty} \left(x^{(i)}\right)^2 < \infty.$$

Let us choose some parameters $\mu > 0$ and $Q_f > 1$, which define the following function

$$f_{\mu,Q_f}(x) = \frac{\mu(Q_f-1)}{8}\left\{(x^{(1)})^2 + \sum_{i=1}^{\infty}(x^{(i)} - x^{(i+1)})^2 - 2x^{(1)}\right\} + \frac{\mu}{2} \parallel x \parallel^2 .$$

Denote

$$A = \begin{pmatrix} 2 & -1 & 0 & 0 \\ -1 & 2 & -1 & 0 \\ 0 & -1 & 2 & \ddots \\ 0 & 0 & \ddots & \ddots \end{pmatrix}.$$

Then $f''(x) = \frac{\mu(Q_f-1)}{4}A + \mu I$, where I is the unit operator in R^∞. In the previous section we have already seen that $0 \preceq A \preceq 4I$. Therefore

$$\mu I \preceq f''(x) \preceq (\mu(Q_f - 1) + \mu)I = \mu Q_f I.$$

This means that $f_{\mu,Q_f} \in \mathcal{S}^{\infty,1}_{\mu,\mu Q_f}(R^\infty)$. Note that the condition number of function f_{μ,Q_f} is

$$Q_{f_{\mu,Q_f}} = \frac{\mu Q_f}{\mu} = Q_f.$$

Let us find the minimum of function $f_{\mu,\mu Q_f}$. The first-order optimality condition

$$f'_{\mu,\mu Q_f}(x) \equiv \left(\frac{\mu(Q_f-1)}{4}A + \mu I\right)x - \frac{\mu(Q_f-1)}{4}e_1 = 0$$

can be written as

$$\left(A + \frac{4}{Q_f-1}\right)x = e_1.$$

The coordinate form of this equation is as follows:

$$\begin{aligned} 2\frac{Q_f+1}{Q_f-1}x^{(1)} - x^{(2)} &= 1, \\ x^{(k+1)} - 2\frac{Q_f+1}{Q_f-1}x^{(k)} + x^{(k-1)} &= 0, \quad k = 2, \ldots . \end{aligned} \tag{2.1.25}$$

Let q be the smallest root of the equation

$$q^2 - 2\frac{Q_f+1}{Q_f-1}q + 1 = 0,$$

that is $q = \frac{\sqrt{Q_f}-1}{\sqrt{Q_f}+1}$. Then the sequence $(x^*)^{(k)} = q^k$, $k = 1, 2, \ldots$, satisfies the system (2.1.25). Thus, we come to the following result.

THEOREM 2.1.13 *For any $x_0 \in R^\infty$ and any constants $\mu > 0$, $Q_f > 1$ there exists a function $f \in S^{\infty,1}_{\mu,\mu Q_f}(R^\infty)$ such that for any first-order method $\mathcal{M}$ satisfying Assumption 2.1.4, we have*

$$\| x_k - x^* \|^2 \geq \left(\frac{\sqrt{Q_f}-1}{\sqrt{Q_f}+1}\right)^{2k} \| x_0 - x^* \|^2,$$

$$f(x_k) - f^* \geq \frac{\mu}{2}\left(\frac{\sqrt{Q_f}-1}{\sqrt{Q_f}+1}\right)^{2k} \| x_0 - x^* \|^2,$$

where x^* is the minimum of function f and $f^* = f(x^*)$.

Proof: Indeed, we can assume that $x_0 = 0$. Let us choose $f(x) = f_{\mu,\mu Q_f}(x)$. Then

$$\| x_0 - x^* \|^2 = \sum_{i=1}^{\infty} [(x^*)^{(i)}]^2 = \sum_{i=1}^{\infty} q^{2i} = \frac{q^2}{1-q^2}.$$

Since $f''_{\mu,\mu Q_f}(x)$ is a three-diagonal operator and $f'_{\mu,\mu Q_f}(0) = e_1$, we conclude that $x_k \in R^{k,\infty}$. Therefore

$$\| x_k - x^* \|^2 \geq \sum_{i=k+1}^{\infty} [(x^*)^{(i)}]^2 = \sum_{i=k+1}^{\infty} q^{2i} = \frac{q^{2(k+1)}}{1-q^2} = q^{2k} \| x_0 - x^* \|^2 .$$

The second bound of the theorem follows from the first one and Theorem 2.1.8. □

2.1.5 Gradient method

Let us check how the gradient method works on the problem

$$\min_{x \in R^n} f(x)$$

with $f \in \mathcal{F}^{1,1}_L(R^n)$. Recall that the scheme of the gradient method is as follows.

Gradient method
0. Choose $x_0 \in R^n$. **1.** kth iteration ($k \geq 0$). a). Compute $f(x_k)$ and $f'(x_k)$. b). Find $x_{k+1} = x_k - h_k f'(x_k)$ (see Section 2 for step-size rules).

In this section we analyze the simplest variant of the gradient scheme with $h_k = h > 0$. It is possible to show that for all other reasonable step-size rules the rate of convergence of this method is similar. Denote by x^* the optimal point of our problem and $f^* = f(x^*)$.

THEOREM 2.1.14 *Let $f \in \mathcal{F}_L^{1,1}(R^n)$ and $0 < h < \frac{2}{L}$. Then the gradient method generates a sequence $\{x_k\}$, which converges as follows:*

$$f(x_k) - f^* \le \frac{2(f(x_0)-f^*)\|x_0-x^*\|^2}{2\|x_0-x^*\|^2+k\cdot h(2-Lh)\cdot(f(x_0)-f^*)}.$$

Proof: Denote $r_k = \| x_k - x^* \|$. Then

$$\begin{aligned} r_{k+1}^2 &= \| x_k - x^* - hf'(x_k) \|^2 \\ &= r_k^2 - 2h\langle f'(x_k), x_k - x^*\rangle + h^2 \| f'(x_k) \|^2 \\ &\le r_k^2 - h(\tfrac{2}{L} - h) \| f'(x_k) \|^2 \end{aligned}$$

(we use (2.1.8) and $f'(x^*) = 0$). Therefore $r_k \le r_0$. In view of (2.1.6) we have

$$\begin{aligned} f(x_{k+1}) &\le f(x_k) + \langle f'(x_k), x_{k+1} - x_k\rangle + \tfrac{L}{2} \| x_{k+1} - x_k \|^2 \\ &= f(x_k) - \omega \| f'(x_k) \|^2, \end{aligned}$$

where $\omega = h(1 - \frac{L}{2}h)$. Denote $\Delta_k = f(x_k) - f^*$. Then

$$\Delta_k \le \langle f'(x_k), x_k - x^*\rangle \le r_0 \| f'(x_k) \| .$$

Therefore $\Delta_{k+1} \le \Delta_k - \frac{\omega}{r_0^2}\Delta_k^2$. Thus,

$$\frac{1}{\Delta_{k+1}} \ge \frac{1}{\Delta_k} + \frac{\omega}{r_0^2} \cdot \frac{\Delta_k}{\Delta_{k+1}} \ge \frac{1}{\Delta_k} + \frac{\omega}{r_0^2}.$$

Summing up these inequalities, we get

$$\frac{1}{\Delta_{k+1}} \ge \frac{1}{\Delta_0} + \frac{\omega}{r_0^2}(k+1).$$

□

In order to choose the optimal step size, we need to maximize the function $\phi(h) = h(2 - Lh)$ with respect to h. The first-order optimality condition $\phi'(h) = 2 - 2Lh = 0$ provides us with the value $h^* = \frac{1}{L}$. In this case we get the following efficiency estimate of the gradient method:

$$f(x_k) - f^* \le \frac{2L(f(x_0)-f^*)\|x_0-x^*\|^2}{2L\|x_0-x^*\|^2+k\cdot(f(x_0)-f^*)}. \quad (2.1.26)$$

Further, in view of (2.1.6) we have

$$\begin{aligned} f(x_0) &\le f^* + \langle f'(x^*), x_0 - x^* \rangle + \tfrac{L}{2} \| x_0 - x^* \|^2 \\ &= f^* + \tfrac{L}{2} \| x_0 - x^* \|^2 . \end{aligned}$$

Since the right-hand side of inequality (2.1.26) is increasing in $f(x_0) - f^*$, we obtain the following result.

COROLLARY 2.1.2 *If $h = \frac{1}{L}$ and $f \in \mathcal{F}_L^{1,1}(R^n)$, then*

$$f(x_k) - f^* \le \frac{2L\|x_0 - x^*\|^2}{k+4}. \qquad (2.1.27)$$

Let us estimate the performance of the gradient method on the class of strongly convex functions.

THEOREM 2.1.15 *If $f \in \mathcal{S}_{\mu,L}^{1,1}(R^n)$ and $0 < h \le \frac{2}{\mu+L}$, then the gradient method generates a sequence $\{x_k\}$ such that*

$$\| x_k - x^* \|^2 \le \left(1 - \frac{2h\mu L}{\mu+L}\right)^k \| x_0 - x^* \|^2 .$$

If $h = \frac{2}{\mu+L}$, then

$$\begin{aligned} \| x_k - x^* \| &\le \left(\frac{Q_f - 1}{Q_f + 1}\right)^k \| x_0 - x^* \|, \\ f(x_k) - f^* &\le \frac{L}{2} \left(\frac{Q_f - 1}{Q_f + 1}\right)^{2k} \| x_0 - x^* \|^2, \end{aligned}$$

where $Q_f = L/\mu$.

Proof: Denote $r_k = \| x_k - x^* \|$. Then

$$\begin{aligned} r_{k+1}^2 &= \| x_k - x^* - hf'(x_k) \|^2 \\ &= r_k^2 - 2h\langle f'(x_k), x_k - x^* \rangle + h^2 \| f'(x_k) \|^2 \\ &\le \left(1 - \frac{2h\mu L}{\mu+L}\right) r_k^2 + h\left(h - \frac{2}{\mu+L}\right) \| f'(x_k) \|^2 \end{aligned}$$

(we use (2.1.24) and $f'(x^*) = 0$). The last inequality in the theorem follows from the previous one and (2.1.6). □

Recall that we have seen already the step-size rule $h = \frac{2}{\mu+L}$ and the linear rate of convergence of the gradient method in Section 1.2.3, Theorem 1.2.4. But that was only a local result.

Comparing the rate of convergence of the gradient method with the lower complexity bounds (Theorems 2.1.7 and 2.1.13), we can see that the gradient method is far from being optimal for classes $\mathcal{F}_L^{1,1}(R^n)$ and $\mathcal{S}_{\mu,L}^{1,1}(R^n)$. We should also note that on these problem classes the standard unconstrained minimization methods (conjugate gradients, variable metric) have a similar global efficiency bound. The optimal methods for minimizing smooth convex and strongly convex functions will be considered in the next section.

2.2 Optimal Methods

(Optimal methods; Convex sets; Constrained minimization problem; Gradient mapping; Minimization methods over a simple set.)

2.2.1 Optimal methods

In this section we consider an unconstrained minimization problem

$$\min_{x \in R^n} f(x),$$

with f being strongly convex: $f \in \mathcal{S}_{\mu,L}^{1,1}(R^n)$, $\mu \geq 0$. Formally, this family of classes contains also the class of convex functions with Lipschitz gradient ($\mathcal{S}_{0,L}^{1,1}(R^n) \equiv \mathcal{F}_L^{1,1}(R^n)$).

In the previous section we proved the following efficiency estimates for the gradient method:

$$\mathcal{F}_L^{1,1}(R^n): \qquad f(x_k) - f^* \leq \frac{2L\|x_0 - x^*\|^2}{k+4},$$

$$\mathcal{S}_{\mu,L}^{1,1}(R^n): \qquad f(x_k) - f^* \leq \frac{L}{2}\left(\frac{L-\mu}{L+\mu}\right)^{2k} \| x_0 - x^* \|^2 .$$

These estimates differ from our lower complexity bounds (Theorem 2.1.7 and Theorem 2.1.13) by an order of magnitude. Of course, in general this does not mean that the gradient method is not optimal since the lower bounds might be too optimistic. However, we will see that in our case the lower bounds are exact up to a constant factor. We prove that by constructing a method that has corresponding efficiency bounds.

Recall that the gradient method forms a relaxation sequence:

$$f(x_{k+1}) \leq f(x_k).$$

This fact is crucial for justification of its convergence rate (Theorem 2.1.14). However, in convex optimization the optimal methods never rely on relaxation. Firstly, for some problem classes this property is too expensive. Secondly, the schemes and efficiency estimates of optimal methods are derived from some *global* topological properties of convex functions. From this point of view, relaxation is a too "microscopic" property to be useful.

The schemes and efficiency bounds of optimal methods are based on the notion of *estimate sequence.*

DEFINITION 2.2.1 *A pair of sequences* $\{\phi_k(x)\}_{k=0}^{\infty}$ *and* $\{\lambda_k\}_{k=0}^{\infty}$, $\lambda_k \geq 0$ *is called an* estimate sequence *of function* $f(x)$ *if*

$$\lambda_k \to 0$$

and for any $x \in R^n$ *and all* $k \geq 0$ *we have*

$$\phi_k(x) \leq (1-\lambda_k)f(x) + \lambda_k \phi_0(x). \tag{2.2.1}$$

The next statement explains why these objects could be useful.

LEMMA 2.2.1 *If for some sequence* $\{x_k\}$ *we have*

$$f(x_k) \leq \phi_k^* \equiv \min_{x \in R^n} \phi_k(x), \tag{2.2.2}$$

then $f(x_k) - f^* \leq \lambda_k[\phi_0(x^*) - f^*] \to 0$.

Proof: Indeed,

$$\begin{aligned} f(x_k) &\leq \phi_k^* = \min_{x \in R^n} \phi_k(x) \leq \min_{x \in R^n}[(1-\lambda_k)f(x) + \lambda_k \phi_0(x)] \\ &\leq (1-\lambda_k)f(x^*) + \lambda_k \phi_0(x^*). \end{aligned}$$

□

Thus, for any sequence $\{x_k\}$, satisfying (2.2.2) we can derive its rate of convergence *directly* from the rate of convergence of sequence $\{\lambda_k\}$. However, at this moment we have two serious questions. Firstly, we do not know how to form an estimate sequence. And secondly, we do not know how we can ensure (2.2.2). The first question is simpler, so let us answer it.

LEMMA 2.2.2 *Assume that:*

1 $f \in \mathcal{S}_{\mu,L}^{1,1}(R^n)$,

2 $\phi_0(x)$ *is an arbitrary function on* R^n,

3 $\{y_k\}_{k=0}^{\infty}$ *is an arbitrary sequence in* R^n,

4 $\{\alpha_k\}_{k=0}^{\infty}: \ \alpha_k \in (0,1), \quad \sum\limits_{k=0}^{\infty} \alpha_k = \infty,$

5 $\lambda_0 = 1.$

Then the pair of sequences $\{\phi_k(x)\}_{k=0}^{\infty}$, $\{\lambda_k\}_{k=0}^{\infty}$ *recursively defined by t:*

$$\begin{aligned} \lambda_{k+1} &= (1-\alpha_k)\lambda_k, \\ \phi_{k+1}(x) &= (1-\alpha_k)\phi_k(x) \qquad\qquad (2.2.3) \\ &\quad +\alpha_k[f(y_k) + \langle f'(y_k), x - y_k\rangle + \tfrac{\mu}{2} \parallel x - y_k \parallel^2], \end{aligned}$$

is an estimate sequence.

Proof: Indeed, $\phi_0(x) \le (1-\lambda_0)f(x) + \lambda_0\phi_0(x) \equiv \phi_0(x)$. Further, let (2.2.1) hold for some $k \ge 0$. Then

$$\begin{aligned} \phi_{k+1}(x) \le \ & (1-\alpha_k)\phi_k(x) + \alpha_k f(x) \\ = \ & (1-(1-\alpha_k)\lambda_k)f(x) + (1-\alpha_k)(\phi_k(x) - (1-\lambda_k)f(x)) \\ \le \ & (1-(1-\alpha_k)\lambda_k)f(x) + (1-\alpha_k)\lambda_k\phi_0(x) \\ = \ & (1-\lambda_{k+1})f(x) + \lambda_{k+1}\phi_0(x). \end{aligned}$$

It remains to note that condition 4) ensures $\lambda_k \to 0$. □

Thus, the above statement provides us with some rules for updating the estimate sequence. Now we have two control sequences, which can help to ensure inequality (2.2.2). Note that we are also free in the choice of initial function $\phi_0(x)$. Let us choose it as a simple quadratic function. Then we can obtain the exact description of the way ϕ_k^* varies.

LEMMA 2.2.3 *Let* $\phi_0(x) = \phi_0^* + \frac{\gamma_0}{2} \parallel x - v_0 \parallel^2$. *Then the process (2.2.3) preserves the canonical form of functions* $\{\phi_k(x)\}$*:*

$$\phi_k(x) \equiv \phi_k^* + \tfrac{\gamma_k}{2} \parallel x - v_k \parallel^2, \qquad (2.2.4)$$

where the sequences $\{\gamma_k\}$, $\{v_k\}$ *and* $\{\phi_k^*\}$ *are defined as follows:*

$$\begin{aligned} \gamma_{k+1} = \ & (1-\alpha_k)\gamma_k + \alpha_k\mu, \\ v_{k+1} = \ & \tfrac{1}{\gamma_{k+1}}[(1-\alpha_k)\gamma_k v_k + \alpha_k\mu y_k - \alpha_k f'(y_k)], \\ \phi_{k+1}^* = \ & (1-\alpha_k)\phi_k + \alpha_k f(y_k) - \tfrac{\alpha_k^2}{2\gamma_{k+1}} \parallel f'(y_k) \parallel^2 \\ & + \tfrac{\alpha_k(1-\alpha_k)\gamma_k}{\gamma_{k+1}} \left(\tfrac{\mu}{2} \parallel y_k - v_k \parallel^2 + \langle f'(y_k), v_k - y_k\rangle \right). \end{aligned}$$

Proof: Note that $\phi_0''(x) = \gamma_0 I_n$. Let us prove that $\phi_k''(x) = \gamma_k I_n$ for all $k \geq 0$. Indeed, if that is true for some k, then

$$\phi_{k+1}''(x) = (1-\alpha_k)\phi_k''(x) + \alpha_k \mu I_n = ((1-\alpha_k)\gamma_k + \alpha_k\mu)I_n \equiv \gamma_{k+1} I_n.$$

This justifies the canonical form (2.2.4) of functions $\phi_k(x)$.

Further,

$$\begin{aligned}\phi_{k+1}(x) &= (1-\alpha_k)\left(\phi_k^* + \tfrac{\gamma_k}{2} \| x - v_k \|^2\right) \\ &\quad + \alpha_k[f(y_k) + \langle f'(y_k), x - y_k\rangle + \tfrac{\mu}{2} \| x - y_k \|^2].\end{aligned}$$

Therefore the equation $\phi_{k+1}'(x) = 0$, which is the first-order optimality condition for function $\phi_{k+1}(x)$, looks as follows:

$$(1-\alpha_k)\gamma_k(x - v_k) + \alpha_k f'(y_k) + \alpha_k\mu(x - y_k) = 0.$$

From that we get the equation for the point v_{k+1}, which is the minimum of the function $\phi_{k+1}(x)$.

Finally, let us compute ϕ_{k+1}^*. In view of the recursion rule for the sequence $\{\phi_k(x)\}$, we have

$$\begin{aligned}&\phi_{k+1}^* + \tfrac{\gamma_{k+1}}{2} \| y_k - v_{k+1} \|^2 = \phi_{k+1}(y_k) \\ &= (1-\alpha_k)\left(\phi_k^* + \tfrac{\gamma_k}{2} \| y_k - v_k \|^2\right) + \alpha_k f(y_k).\end{aligned} \tag{2.2.5}$$

Note that in view of the relation for v_{k+1},

$$v_{k+1} - y_k = \tfrac{1}{\gamma_{k+1}}[(1-\alpha_k)\gamma_k(v_k - y_k) - \alpha_k f'(y_k)].$$

Therefore

$$\begin{aligned}\tfrac{\gamma_{k+1}}{2} \| v_{k+1} - y_k \|^2 &= \tfrac{1}{2\gamma_{k+1}}[(1-\alpha_k)^2\gamma_k^2 \| v_k - y_k \|^2 \\ &\quad -2\alpha_k(1-\alpha_k)\gamma_k\langle f'(y_k), v_k - y_k\rangle \\ &\quad +\alpha_k^2 \| f'(y_k) \|^2].\end{aligned}$$

It remains to substitute this relation into (2.2.5) noting that the factor for the term $\| y_k - v_k \|^2$ in this expression is as follows:

$$\begin{aligned}(1-\alpha_k)\tfrac{\gamma_k}{2} - \tfrac{1}{2\gamma_{k+1}}(1-\alpha_k)^2\gamma_k^2 &= (1-\alpha_k)\tfrac{\gamma_k}{2}\left(1 - \tfrac{(1-\alpha_k)\gamma_k}{\gamma_{k+1}}\right) \\ &= (1-\alpha_k)\tfrac{\gamma_k}{2} \cdot \tfrac{\alpha_k\mu}{\gamma_{k+1}}.\end{aligned}$$

□

Now the situation is more clear and we are close to getting an algorithmic scheme. Indeed, assume that we already have x_k:

$$\phi_k^* \geq f(x_k).$$

Then, in view of the previous lemma,

$$\begin{aligned}\phi_{k+1}^* \geq \ & (1-\alpha_k)f(x_k) + \alpha_k f(y_k) - \tfrac{\alpha_k^2}{2\gamma_{k+1}} \parallel f'(y_k) \parallel^2 \\ & + \tfrac{\alpha_k(1-\alpha_k)\gamma_k}{\gamma_{k+1}} \langle f'(y_k), v_k - y_k \rangle.\end{aligned}$$

Since $f(x_k) \geq f(y_k) + \langle f'(y_k), x_k - y_k \rangle$, we get the following estimate:

$$\begin{aligned}\phi_{k+1}^* \ \geq \ & f(y_k) - \tfrac{\alpha_k^2}{2\gamma_{k+1}} \parallel f'(y_k) \parallel^2 \\ & +(1-\alpha_k)\langle f'(y_k), \tfrac{\alpha_k\gamma_k}{\gamma_{k+1}}(v_k - y_k) + x_k - y_k \rangle.\end{aligned}$$

Let us look at this inequality. We want to have $\phi_{k+1}^* \geq f(x_{k+1})$. Recall, that we can ensure the inequality

$$f(y_k) - \tfrac{1}{2L} \parallel f'(y_k) \parallel^2 \geq f(x_{k+1})$$

in many different ways. The simplest one is just to take the gradient step

$$x_{k+1} = y_k - h_k f'(x_k)$$

with $h_k = \frac{1}{L}$ (see (2.1.6)). Let us define α_k as follows:

$$L\alpha_k^2 = (1-\alpha_k)\gamma_k + \alpha_k\mu \quad (= \gamma_{k+1}).$$

Then $\frac{\alpha_k^2}{2\gamma_{k+1}} = \frac{1}{2L}$ and we can replace the previous inequality by the following:

$$\phi_{k+1}^* \geq f(x_{k+1}) + (1-\alpha_k)\langle f'(y_k), \tfrac{\alpha_k\gamma_k}{\gamma_{k+1}}(v_k - y_k) + x_k - y_k \rangle.$$

Now we can use our freedom in the choice of y_k. Let us find it from the equation:

$$\tfrac{\alpha_k\gamma_k}{\gamma_{k+1}}(v_k - y_k) + x_k - y_k = 0.$$

That is

$$y_k = \tfrac{\alpha_k\gamma_k v_k + \gamma_{k+1}x_k}{\gamma_k + \alpha_k\mu}.$$

Thus, we come to the following method.

General scheme of optimal method

(2.2.6)

0. Choose $x_0 \in R^n$ and $\gamma_0 > 0$. Set $v_0 = x_0$.

1. kth iteration ($k \geq 0$).

a). Compute $\alpha_k \in (0,1)$ from equation

$$L\alpha_k^2 = (1-\alpha_k)\gamma_k + \alpha_k\mu.$$

Set $\gamma_{k+1} = (1-\alpha_k)\gamma_k + \alpha_k\mu$.

b). Choose

$$y_k = \frac{\alpha_k\gamma_k v_k + \gamma_{k+1}x_k}{\gamma_k + \alpha_k\mu}$$

and compute $f(y_k)$ and $f'(y_k)$.

c). Find x_{k+1} such that

$$f(x_{k+1}) \leq f(y_k) - \frac{1}{2L} \parallel f'(y_k) \parallel^2$$

(see Section 1.2.3 for the step-size rules).

d). Set $v_{k+1} = \frac{(1-\alpha_k)\gamma_k v_k + \alpha_k\mu y_k - \alpha_k f'(y_k)}{\gamma_{k+1}}$.

Note that in Step 1c) of this scheme we can choose any x_{k+1} satisfying the inequality

$$f(x_{k+1}) \leq f(y_k) - \frac{\omega}{2} \parallel f'(y_k) \parallel^2$$

with some $\omega > 0$. Then the constant $\frac{1}{\omega}$ replaces L in the equation of Step 1a).

THEOREM 2.2.1 *The scheme (2.2.6) generates a sequence $\{x_k\}_{k=0}^{\infty}$ such that*

$$f(x_k) - f^* \leq \lambda_k \left[f(x_0) - f^* + \frac{\gamma_0}{2} \parallel x_0 - x^* \parallel^2\right],$$

where $\lambda_0 = 1$ and $\lambda_k = \Pi_{i=0}^{k-1}(1-\alpha_i)$.

Proof: Indeed, let us choose $\phi_0(x) = f(x_0) + \frac{\gamma_0}{2} \parallel x - v_0 \parallel^2$. Then $f(x_0) = \phi_0^*$ and we get $f(x_k) \leq \phi_k^*$ by construction of the scheme. It remains to use Lemma 2.2.1. □

Thus, in order to estimate the rate of convergence of (2.2.6), we need to understand how fast λ_k goes to zero.

LEMMA 2.2.4 *If in the scheme (2.2.6) $\gamma_0 \geq \mu$, then*

$$\lambda_k \leq \min\left\{\left(1-\sqrt{\tfrac{\mu}{L}}\right)^k, \tfrac{4L}{(2\sqrt{L}+k\sqrt{\gamma_0})^2}\right\}. \qquad (2.2.7)$$

Proof: Indeed, if $\gamma_k \geq \mu$, then $\gamma_{k+1} = L\alpha_k^2 = (1-\alpha_k)\gamma_k + \alpha_k\mu \geq \mu$. Since $\gamma_0 \geq \mu$, we conclude that this inequality is valid for all γ_k. Hence, $\alpha_k \geq \sqrt{\frac{\mu}{L}}$ and we have proved the first inequality in (2.2.7).

Further, let us prove that $\gamma_k \geq \gamma_0\lambda_k$. Indeed, since $\gamma_0 = \gamma_0\lambda_0$, we can use induction:

$$\gamma_{k+1} \geq (1-\alpha_k)\gamma_k \geq (1-\alpha_k)\gamma_0\lambda_k = \gamma_0\lambda_{k+1}.$$

Therefore $L\alpha_k^2 = \gamma_{k+1} \geq \gamma_0\lambda_{k+1}$.

Denote $a_k = \frac{1}{\sqrt{\lambda_k}}$. Since $\{\lambda_k\}$ is a decreasing sequence, we have

$$\begin{aligned} a_{k+1} - a_k &= \tfrac{\sqrt{\lambda_k}-\sqrt{\lambda_{k+1}}}{\sqrt{\lambda_k\lambda_{k+1}}} = \tfrac{\lambda_k-\lambda_{k+1}}{\sqrt{\lambda_k\lambda_{k+1}}(\sqrt{\lambda_k}+\sqrt{\lambda_{k+1}})} \\ &\geq \tfrac{\lambda_k-\lambda_{k+1}}{2\lambda_k\sqrt{\lambda_{k+1}}} = \tfrac{\lambda_k-(1-\alpha_k)\lambda_k}{2\lambda_k\sqrt{\lambda_{k+1}}} = \tfrac{\alpha_k}{2\sqrt{\lambda_{k+1}}} \geq \tfrac{1}{2}\sqrt{\tfrac{\gamma_0}{L}}. \end{aligned}$$

Thus, $a_k \geq 1 + \frac{k}{2}\sqrt{\frac{\gamma_0}{L}}$ and the lemma is proved. □

Let us present an exact statement on optimality of (2.2.6).

THEOREM 2.2.2 *Let us take in (2.2.6) $\gamma_0 = L$. Then this scheme generates a sequence $\{x_k\}_{k=0}^{\infty}$ such that*

$$f(x_k) - f^* \leq L\min\left\{\left(1-\sqrt{\tfrac{\mu}{L}}\right)^k, \tfrac{4}{(k+2)^2}\right\} \parallel x_0 - x^* \parallel^2.$$

This means that (2.2.6) is optimal *for unconstrained minimization of the functions from $\mathcal{S}^{1,1}_{\mu,L}(R^n)$, $\mu \geq 0$.*

Proof: We get the above inequality using $f(x_0) - f^* \leq \frac{L}{2} \parallel x_0 - x^* \parallel^2$ and Theorem 2.2.1 with Lemma 2.2.4.

Let $\mu > 0$. From the lower complexity bounds for the class (see Theorem 2.1.13) we have

$$f(x_k) - f^* \geq \tfrac{\mu}{2}\left(\tfrac{\sqrt{Q_f}-1}{\sqrt{Q_f}+1}\right)^{2k} R^2 \geq \tfrac{\mu}{2}\exp\left(-\tfrac{4k}{\sqrt{Q_f}-1}\right) R^2,$$

where $Q_f = L/\mu$ and $R = \| x_0 - x^* \|$. Therefore, the worst case bound for finding x_k satisfying $f(x_k) - f^* \leq \epsilon$ cannot be better than

$$k \geq \frac{\sqrt{Q_f}-1}{4}\left[\ln \tfrac{1}{\epsilon} + \ln \tfrac{\mu}{2} + 2\ln R\right].$$

For our scheme we have

$$f(x_k) - f^* \leq LR^2\left(1 - \sqrt{\tfrac{\mu}{L}}\right)^k \leq LR^2 \exp\left(-\tfrac{k}{\sqrt{Q_f}}\right).$$

Therefore we guarantee that $k \leq \sqrt{Q_f}\left[\ln \frac{1}{\epsilon} + \ln L + 2\ln R\right]$. Thus, the main term in this estimate, $\sqrt{Q_f}\ln\frac{1}{\epsilon}$, is proportional to the lower bound. The same reasoning can be used for the class $\mathcal{S}_{0,L}^{1,1}(R^n)$. □

Let us analyze a variant of the scheme (2.2.6), which uses the gradient step for finding the point x_{k+1}.

Constant Step Scheme, I

0. Choose $x_0 \in R^n$ and $\gamma_0 > 0$. Set $v_0 = x_0$.

1. kth iteration ($k \geq 0$).

a). Compute $\alpha_k \in (0,1)$ from the equation

$$L\alpha_k^2 = (1-\alpha_k)\gamma_k + \alpha_k\mu. \tag{2.2.8}$$

Set $\gamma_{k+1} = (1-\alpha_k)\gamma_k + \alpha_k\mu$.

b). Choose $y_k = \frac{\alpha_k\gamma_k v_k + \gamma_{k+1}x_k}{\gamma_k + \alpha_k\mu}$. Compute $f(y_k)$ and $f'(y_k)$.

c). Set $x_{k+1} = y_k - \frac{1}{L}f'(y_k)$ and

$$v_{k+1} = \frac{1}{\gamma_{k+1}}[(1-\alpha_k)\gamma_k v_k + \alpha_k\mu y_k - \alpha_k f'(y_k)].$$

Let us demonstrate that this scheme can be rewritten in a simpler form. Note that

$$\begin{aligned} y_k &= \frac{1}{\gamma_k+\alpha_k\mu}(\alpha_k\gamma_k v_k + \gamma_{k+1}x_k), \\ x_{k+1} &= y_k - \tfrac{1}{L}f'(y_k), \\ v_{k+1} &= \frac{1}{\gamma_{k+1}}[(1-\alpha_k)\gamma_k v_k + \alpha_k\mu y_k - \alpha_k f'(y_k)]. \end{aligned}$$

Therefore

$$\begin{aligned} v_{k+1} &= \tfrac{1}{\gamma_{k+1}}\{\tfrac{(1-\alpha_k)}{\alpha_k}[(\gamma_k+\alpha_k\mu)y_k-\gamma_{k+1}x_k]+\alpha_k\mu y_k-\alpha_k f'(y_k)\} \\ &= \tfrac{1}{\gamma_{k+1}}\left\{\tfrac{(1-\alpha_k)\gamma_k}{\alpha_k}y_k+\mu y_k\right\}-\tfrac{1-\alpha_k}{\alpha_k}x_k-\tfrac{\alpha_k}{\gamma_{k+1}}f'(y_k) \\ &= x_k+\tfrac{1}{\alpha_k}(y_k-x_k)-\tfrac{1}{\alpha_k L}f'(y_k) \\ &= x_k+\tfrac{1}{\alpha_k}(x_{k+1}-x_k). \end{aligned}$$

Hence,

$$\begin{aligned} y_{k+1} &= \tfrac{1}{\gamma_{k+1}+\alpha_{k+1}\mu}(\alpha_{k+1}\gamma_{k+1}v_{k+1}+\gamma_{k+2}x_{k+1}) \\ &= x_{k+1}+\tfrac{\alpha_{k+1}\gamma_{k+1}(v_{k+1}-x_{k+1})}{\gamma_{k+1}+\alpha_{k+1}\mu}=x_{k+1}+\beta_k(x_{k+1}-x_k), \end{aligned}$$

where

$$\beta_k=\tfrac{\alpha_{k+1}\gamma_{k+1}(1-\alpha_k)}{\alpha_k(\gamma_{k+1}+\alpha_{k+1}\mu)}.$$

Thus, we managed to get rid of $\{v_k\}$. Let us do the same with γ_k. We have

$$\alpha_k^2 L=(1-\alpha_k)\gamma_k+\mu\alpha_k\equiv\gamma_{k+1}.$$

Therefore

$$\begin{aligned} \beta_k &= \tfrac{\alpha_{k+1}\gamma_{k+1}(1-\alpha_k)}{\alpha_k(\gamma_{k+1}+\alpha_{k+1}\mu)}=\tfrac{\alpha_{k+1}\gamma_{k+1}(1-\alpha_k)}{\alpha_k(\gamma_{k+1}+\alpha_{k+1}^2L-(1-\alpha_{k+1})\gamma_{k+1})} \\ &= \tfrac{\gamma_{k+1}(1-\alpha_k)}{\alpha_k(\gamma_{k+1}+\alpha_{k+1}L)}=\tfrac{\alpha_k(1-\alpha_k)}{\alpha_k^2+\alpha_{k+1}}. \end{aligned}$$

Note also that $\alpha_{k+1}^2=(1-\alpha_{k+1})\alpha_k^2+q\alpha_{k+1}$ with $q=\mu/L$, and

$$\alpha_0^2L=(1-\alpha_0)\gamma_0+\mu\alpha_0.$$

The latter relation means that γ_0 can be seen as a function of α_0. Thus, we can completely eliminate the sequence $\{\gamma_k\}$. Let us write down the corresponding scheme.

Constant Step Scheme, II

0. Choose $x_0 \in R^n$ and $\alpha_0 \in (0,1)$.
Set $y_0 = x_0$ and $q = \frac{\mu}{L}$.

1. kth iteration ($k \geq 0$).

a). Compute $f(y_k)$ and $f'(y_k)$. Set

$$x_{k+1} = y_k - \tfrac{1}{L} f'(y_k). \tag{2.2.9}$$

b). Compute $\alpha_{k+1} \in (0,1)$ from equation

$$\alpha_{k+1}^2 = (1-\alpha_{k+1})\alpha_k^2 + q\alpha_{k+1},$$

and set $\beta_k = \frac{\alpha_k(1-\alpha_k)}{\alpha_k^2+\alpha_{k+1}}$,

$$y_{k+1} = x_{k+1} + \beta_k(x_{k+1} - x_k).$$

The rate of convergence of the above scheme can be derived from Theorem 2.2.1 and Lemma 2.2.4. Let us write down the corresponding statement in terms of α_0.

THEOREM 2.2.3 *If in scheme (2.2.9)*

$$\alpha_0 \geq \sqrt{\tfrac{\mu}{L}}, \tag{2.2.10}$$

then

$$f(x_k) - f^* \leq \min \left\{ \left(1 - \sqrt{\tfrac{\mu}{L}}\right)^k, \frac{4L}{(2\sqrt{L}+k\sqrt{\gamma_0})^2} \right\}$$

$$\times \left[f(x_0) - f^* + \tfrac{\gamma_0}{2} \parallel x_0 - x^* \parallel^2 \right],$$

where $\gamma_0 = \frac{\alpha_0(\alpha_0 L - \mu)}{1-\alpha_0}$.

We do not need to prove this theorem since the initial scheme is not changed. We change only notation. In Theorem 2.2.3 condition (2.2.10) is equivalent to $\gamma_0 \geq \mu$.

Scheme (2.2.9) becomes very simple if we choose $\alpha_0 = \sqrt{\frac{\mu}{L}}$ (this corresponds to $\gamma_0 = \mu$). Then

$$\alpha_k = \sqrt{\tfrac{\mu}{L}}, \quad \beta_k = \frac{\sqrt{L}-\sqrt{\mu}}{\sqrt{L}+\sqrt{\mu}}$$

for all $k \geq 0$. Thus, we come to the following process.

Constant step scheme, III	
0. Choose $y_0 = x_0 \in R^n$.	
1. kth iteration ($k \geq 0$). $$x_{k+1} = y_k - \tfrac{1}{L} f'(y_k),$$ $$y_{k+1} = x_{k+1} + \tfrac{\sqrt{L}-\sqrt{\mu}}{\sqrt{L}+\sqrt{\mu}}(x_{k+1} - x_k).$$	(2.2.11)

However, note that this process does not work for $\mu = 0$. The choice $\gamma_0 = L$ (which changes corresponding value of α_0) is safer.

2.2.2 Convex sets

Let us try to understand which *constrained* minimization problem we can solve. Let us start from the simplest problem of this type, the problem without functional constraints:

$$\min_{x \in Q} f(x),$$

where Q is some subset of R^n. In order to make our problem tractable, we should impose some assumptions on the set Q. And first of all, let us answer the following question: Which sets fit naturally the class of convex functions? From definition of convex function,

$$f(\alpha x + (1-\alpha)y) \leq \alpha f(x) + (1-\alpha) f(y), \quad \forall x, y \in R^n, \ \alpha \in [0,1],$$

we see that it is implicitly assumed that it is possible to check this inequality at any point of the *segment* $[x, y]$:

$$[x, y] = \{z = \alpha x + (1-\alpha)y, \ \alpha \in [0,1]\}.$$

Thus, it would be natural to consider a set that contains the whole segment $[x, y]$ provided that the end points x and y belong to the set. Such sets are called *convex*.

DEFINITION 2.2.2 *Set Q is called* convex *if for any x, $y \in Q$ and α from $[0,1]$ we have*

$$\alpha x + (1-\alpha)y \in Q.$$

The point $\alpha x + (1-\alpha)y$ with $\alpha \in [0,1]$ is called a *convex combination* of these two points.

In fact, we have already met some convex sets.

LEMMA 2.2.5 *If $f(x)$ is a convex function, then for any $\beta \in R^1$ its* level set

$$\mathcal{L}_f(\beta) = \{x \in R^n \mid f(x) \leq \beta\}$$

is either convex or empty.

Proof: Indeed, let x and y belong to $\mathcal{L}_f(\beta)$. Then $f(x) \leq \beta$ and $f(y) \leq \beta$. Therefore

$$f(\alpha x + (1-\alpha)y) \leq \alpha f(x) + (1-\alpha)f(y) \leq \beta.$$

□

LEMMA 2.2.6 *Let $f(x)$ be a convex function. Then its* epigraph

$$\mathcal{E}_f = \{(x,\tau) \in R^{n+1} \mid f(x) \leq \tau\}$$

is a convex set.

Proof: Indeed, let $z_1 = (x_1, \tau_1) \in \mathcal{E}_f$ and $z_2 = (x_2, \tau_2) \in \mathcal{E}_f$. Then for any $\alpha \in [0,1]$ we have

$$z_\alpha \equiv \alpha z_1 + (1-\alpha)z_2 = (\alpha x_1 + (1-\alpha)x_2, \alpha\tau_1 + (1-\alpha)\tau_2),$$

$$f(\alpha x_1 + (1-\alpha)x_2) \leq \alpha f(x_1) + (1-\alpha)f(x_2) \leq \alpha\tau_1 + (1-\alpha)\tau_2.$$

Thus, $z_\alpha \in \mathcal{E}_f$. □

Let us look at some properties of convex sets.

THEOREM 2.2.4 *Let $Q_1 \subseteq R^n$ and $Q_2 \subseteq R^m$ be convex sets and $\mathcal{A}(x)$ be a linear operator:*

$$\mathcal{A}(x) = Ax + b: \ R^n \ \to \ R^m.$$

Then all sets below are convex:

1. *Intersection ($m = n$): $Q_1 \bigcap Q_2 = \{x \in R^n \mid x \in Q_1, \ x \in Q_2\}$.*
2. *Sum ($m = n$): $Q_1 + Q_2 = \{z = x + y \mid x \in Q_1, \ y \in Q_2\}$.*
3. *Direct sum: $Q_1 \times Q_2 = \{(x,y) \in R^{n+m} \mid x \in Q_1, \ y \in Q_2\}$.*
4. *Conic hull: $\mathcal{K}(Q_1) = \{z \in R^n \mid z = \beta x, \ x \in Q_1, \beta \geq 0\}$.*

5. *Convex hull*

$$\text{Conv}\,(Q_1, Q_2) = \{z \in R^n \mid \quad z = \alpha x + (1-\alpha),$$

$$y, x \in Q_1,\ y \in Q_2,\ \alpha \in [0,1]\}.$$

6. *Affine image:* $\mathcal{A}(Q_1) = \{y \in R^m \mid y = \mathcal{A}(x),\ x \in Q_1\}$.

7. *Inverse affine image:* $\mathcal{A}^{-1}(Q_2) = \{x \in R^n \mid \mathcal{A}(x) \in Q_2\}$.

Proof: 1. If $x_1 \in Q_1 \bigcap Q_2$, $x_1 \in Q_1 \bigcap Q_2$, then $[x_1, x_2] \subset Q_1$ and $[x_1, x_2] \subset Q_2$. Therefore $[x_1, x_2] \subset Q_1 \bigcap Q_2$.

2. If $z_1 = x_1 + x_2$, $x_1 \in Q_1$, $x_2 \in Q_2$ and $z_2 = y_1 + y_2$, $y_1 \in Q_1$, $y_2 \in Q_2$, then

$$\alpha z_1 + (1-\alpha) z_2 = (\alpha x_1 + (1-\alpha) y_1)_1 + (\alpha x_2 + (1-\alpha) y_2)_2,$$

where $(\cdot)_1 \in Q_1$ and $(\cdot)_2 \in Q_2$.

3. If $z_1 = (x_1, x_2)$, $x_1 \in Q_1$, $x_2 \in Q_2$ and $z_2 = (y_1, y_2)$, $y_1 \in Q_1$, $y_2 \in Q_2$, then

$$\alpha z_1 + (1-\alpha) z_2 = ((\alpha x_1 + (1-\alpha) y_1)_1, (\alpha x_2 + (1-\alpha) y_2)_2),$$

where $(\cdot)_1 \in Q_1$ and $(\cdot)_2 \in Q_2$.

4. If $z_1 = \beta_1 x_1$, $x_1 \in Q_1$, $\beta_1 \geq 0$, and $z_2 = \beta_2 x_2$, $x_2 \in Q_1$, $\beta_2 \geq 0$, then for any $\alpha \in [0,1]$ we have

$$\alpha z_1 + (1-\alpha) z_2 = \alpha \beta_1 x_1 + (1-\alpha) \beta_2 x_2 = \gamma(\bar{\alpha} x_1 + (1-\bar{\alpha}) x_2),$$

where $\gamma = \alpha\beta_1 + (1-\alpha)\beta_2$, and $\bar{\alpha} = \alpha\beta_1/\gamma \in [0,1]$.

5. If $z_1 = \beta_1 x_1 + (1-\beta_1) x_2$, $x_1 \in Q_1$, $x_2 \in Q_2$, $\beta_1 \in [0,1]$, and $z_2 = \beta_2 y_1 + (1-\beta_2) y_2$, $y_1 \in Q_1$, $y_2 \in Q_2$, $\beta_2 \in [0,1]$, then for any $\alpha \in [0,1]$ we have

$$\begin{aligned} \alpha z_1 + (1-\alpha) z_2 &= \alpha(\beta_1 x_1 + (1-\beta_1) x_2) \\ &\quad + (1-\alpha)(\beta_2 y_1 + (1-\beta_2) y_2) \\ &= \bar{\alpha}(\bar{\beta}_1 x_1 + (1-\bar{\beta}_1) y_1) \\ &\quad + (1-\bar{\alpha})(\bar{\beta}_2 x_2 + (1-\bar{\beta}_2) y_2), \end{aligned}$$

where $\bar{\alpha} = \alpha\beta_1 + (1-\alpha)\beta_2$ and $\bar{\beta}_1 = \alpha\beta_1/\bar{\alpha}$, $\bar{\beta}_2 = \alpha(1-\beta_1)/(1-\bar{\alpha})$.

6. If y_1, $y_2 \in \mathcal{A}(Q_1)$ then $y_1 = Ax_1 + b$ and $y_2 = Ax_2 + b$ for some x_1, $x_2 \in Q_1$. Therefore, for $y(\alpha) = \alpha y_1 + (1-\alpha) y_2$, $0 \leq \alpha \leq 1$, we have

$$y(\alpha) = \alpha(Ax_1 + b) + (1-\alpha)(Ax_2 + b) = A(\alpha x_1 + (1-\alpha) x_2) + b.$$

Thus, $y(\alpha) \in \mathcal{A}(Q_1)$.

7. If $x_1, x_2 \in \mathcal{A}^{-1}(Q_2)$ then $Ax_1 + b = y_1$ and $Ax_2 + b = y_2$ for some $y_1, y_2 \in Q_2$. Therefore, for $x(\alpha) = \alpha x_1 + (1-\alpha)x_2$, $0 \le \alpha \le 1$, we have

$$\begin{aligned} \mathcal{A}(x(\alpha)) &= A(\alpha x_1 + (1-\alpha)x_2) + b \\ &= \alpha(Ax_1 + b) + (1-\alpha)(Ax_2 + b) = \alpha y_1 + (1-\alpha)y_2 \in Q_2. \end{aligned}$$

□

Let us give several examples of convex sets.

EXAMPLE 2.2.1 1. *Half-space* $\{x \in R^n \mid \langle a, x\rangle \le \beta\}$ is convex since linear function is convex.

2. *Polytope* $\{x \in R^n \mid \langle a_i, x\rangle \le b_i,\ i = 1 \dots m\}$ is convex as an intersection of convex sets.

3. *Ellipsoid.* Let $A = A^T \succeq 0$. Then the set $\{x \in R^n \mid \langle Ax, x\rangle \le r^2\}$ is convex since function $\langle Ax, x\rangle$ is convex. □

Let us write down the *optimality conditions* for the problem

$$\min_{x \in Q} f(x), \quad f \in \mathcal{F}^1(R^n), \tag{2.2.12}$$

where Q is a closed convex set. It is clear that the old condition

$$f'(x) = 0$$

does not work here.

EXAMPLE 2.2.2 Consider the one-dimensional problem:

$$\min_{x \ge 0} x.$$

Here $x \in R^1$, $Q = \{x : x \ge 0\}$ and $f(x) = x$. Note that $x^* = 0$ but $f'(x^*) = 1 > 0$. □

THEOREM 2.2.5 *Let $f \in \mathcal{F}^1(R^n)$ and Q be a closed convex set. The point x^* is a solution of (2.2.12) if and only if*

$$\langle f'(x^*), x - x^*\rangle \ge 0 \tag{2.2.13}$$

for all $x \in Q$.

Proof: Indeed, if (2.2.13) is true, then

$$f(x) \geq f(x^*) + \langle f'(x^*), x - x^* \rangle \geq f(x^*)$$

for all $x \in Q$.

Let x^* be a solution to (2.2.12). Assume that there exists some $x \in Q$ such that

$$\langle f'(x^*), x - x^* \rangle < 0.$$

Consider the function $\phi(\alpha) = f(x^* + \alpha(x - x^*))$, $\alpha \in [0, 1]$. Note that

$$\phi(0) = f(x^*), \quad \phi'(0) = \langle f'(x^*), x - x^* \rangle < 0.$$

Therefore, for α small enough we have

$$f(x^* + \alpha(x - x^*)) = \phi(\alpha) < \phi(0) = f(x^*).$$

That is a contradiction. □

THEOREM 2.2.6 *Let* $f \in \mathcal{S}^1_\mu(R^n)$ *and* Q *be a closed convex set. Then there exists a unique solution* x^* *of problem (2.2.12).*

Proof: Let $x_0 \in Q$. Consider the set $\bar{Q} = \{x \in Q \mid f(x) \leq f(x_0)\}$. Note that problem (2.2.12) is equivalent to

$$\min\{f(x) \mid x \in \bar{Q}\}. \tag{2.2.14}$$

However, $\bar{Q}$ is bounded: for all $x \in \bar{Q}$ we have

$$f(x_0) \geq f(x) \geq f(x_0) + \langle f'(x_0), x - x_0 \rangle + \tfrac{\mu}{2} \| x - x_0 \|^2 .$$

Hence, $\| x - x_0 \| \leq \frac{2}{\mu} \| f'(x_0) \|$.

Thus, the solution x^* of (2.2.14) ($\equiv$ (2.2.12)) exists. Let us prove that it is unique. Indeed, if x_1^* is also a solution to (2.2.12), then

$$\begin{aligned} f^* &= f(x_1^*) \geq f(x^*) + \langle f'(x^*), x_1^* - x^* \rangle + \tfrac{\mu}{2} \| x_1^* - x^* \|^2 \\ &\geq f^* + \tfrac{\mu}{2} \| x_1^* - x^* \|^2 \end{aligned}$$

(we have used Theorem 2.2.5). Therefore $x_1^* = x^*$. □

2.2.3 Gradient mapping

In the constrained minimization problem the gradient of the objective function should be treated differently as compared to the unconstrained situation. In the previous section we have already seen that its role in optimality conditions is changing. Moreover, we cannot use it anymore in a gradient step since the result could be infeasible, etc. If we look at the main properties of the gradient, which we have used for $f \in \mathcal{F}_L^{1,1}(R^n)$, we can see that two of them are of the most importance. The first one is that the gradient step decreases the function value by an amount comparable with the squared norm of the gradient:

$$f(x - \tfrac{1}{L} f'(x)) \leq f(x) - \tfrac{1}{2L} \parallel f'(x) \parallel^2 .$$

And the second one is the inequality

$$\langle f'(x), x - x^* \rangle \geq \tfrac{1}{L} \parallel f'(x) \parallel^2 .$$

It turns out that for constrained minimization problems we can introduce an object that inherits the most important properties of the gradient.

DEFINITION 2.2.3 *Let us fix some $\gamma > 0$. Denote*

$$\begin{aligned} x_Q(\bar{x}; \gamma) &= \arg\min_{x \in Q} \left[f(\bar{x}) + \langle f'(\bar{x}), x - \bar{x} \rangle + \tfrac{\gamma}{2} \parallel x - \bar{x} \parallel^2 \right], \\ g_Q(\bar{x}; \gamma) &= \gamma(\bar{x} - x_Q(\bar{x}; \gamma)). \end{aligned}$$

We call $g_Q(\gamma, x)$ the gradient mapping *of f on Q.*

For $Q \equiv R^n$ we have

$$x_Q(\bar{x}; \gamma) = \bar{x} - \tfrac{1}{\gamma} f'(\bar{x}), \quad g_Q(\bar{x}; \gamma) = f'(\bar{x}).$$

Thus, the value $\frac{1}{\gamma}$ can be seen as a step size for the "gradient" step

$$\bar{x} \to x_Q(\bar{x}; \gamma).$$

Note that the gradient mapping is well defined in view of Theorem 2.2.6. Moreover, it is defined for all $\bar{x} \in R^n$, not necessarily from Q.

Let us write down the main property of gradient mapping.

THEOREM 2.2.7 *Let $f \in \mathcal{S}_{\mu,L}^{1,1}(R^n)$, $\gamma \geq L$ and $\bar{x} \in R^n$. Then for any $x \in Q$ we have*

$$\begin{aligned} f(x) \geq\ & f(x_Q(\bar{x}; \gamma)) + \langle g_Q(\bar{x}; \gamma), x - \bar{x} \rangle \\ & + \tfrac{1}{2\gamma} \parallel g_Q(\bar{x}; \gamma) \parallel^2 + \tfrac{\mu}{2} \parallel x - \bar{x} \parallel^2 . \end{aligned} \qquad (2.2.15)$$

Proof: Denote $x_Q = x_Q(\gamma, \bar{x})$, $g_Q = g_Q(\gamma, \bar{x})$ and let

$$\phi(x) = f(\bar{x}) + \langle f'(\bar{x}), x - \bar{x}\rangle + \frac{\gamma}{2} \| x - \bar{x} \|^2 .$$

Then $\phi'(x) = f'(\bar{x}) + \gamma(x - \bar{x})$, and for any $x \in Q$ we have

$$\langle f'(\bar{x}) - g_Q, x - x_Q\rangle = \langle \phi'(x_Q), x - x_Q\rangle \geq 0.$$

Hence,

$$\begin{aligned}
f(x) - \frac{\mu}{2} \| x - \bar{x} \|^2 &\geq f(\bar{x}) + \langle f'(\bar{x}), x - \bar{x}\rangle \\
&= f(\bar{x}) + \langle f'(\bar{x}), x_Q - \bar{x}\rangle + \langle f'(\bar{x}), x - x_Q\rangle \\
&\geq f(\bar{x}) + \langle f'(\bar{x}), x_Q - \bar{x}\rangle + \langle g_Q, x - x_Q\rangle \\
&= \phi(x_Q) - \frac{\gamma}{2} \| x_Q - \bar{x} \|^2 + \langle g_Q, x - x_Q\rangle \\
&= \phi(x_Q) - \frac{1}{2\gamma} \| g_Q \|^2 + \langle g_Q, x - x_Q\rangle \\
&= \phi(x_Q) + \frac{1}{2\gamma} \| g_Q \|^2 + \langle g_Q, x - \bar{x}\rangle,
\end{aligned}$$

and $\phi(x_Q) \geq f(x_Q)$ since $\gamma \geq L$. □

COROLLARY 2.2.1 *Let $f \in \mathcal{S}^{1,1}_{\mu,L}(R^n)$, $\gamma \geq L$ and $\bar{x} \in R^n$. Then*

$$f(x_Q(\bar{x};\gamma)) \leq f(\bar{x}) - \frac{1}{2\gamma} \| g_Q(\bar{x};\gamma) \|^2, \qquad (2.2.16)$$

$$\langle g_Q(\bar{x};\gamma), \bar{x} - x^*\rangle \geq \frac{1}{2\gamma} \| g_Q(\bar{x};\gamma) \|^2 + \frac{\mu}{2} \| x - \bar{x} \|^2 . \qquad (2.2.17)$$

Proof: Indeed, using (2.2.15) with $x = \bar{x}$, we get (2.2.16). Using (2.2.15) with $x = x^*$, we get (2.2.17) since $f(x_Q(\bar{x};\gamma)) \geq f(x^*)$. □

2.2.4 Minimization methods for simple sets

Let us show how we can use the gradient mapping for solving the following problem:

$$\min_{x \in Q} f(x),$$

where $f \in \mathcal{S}^{1,1}_{\mu,L}(R^n)$ and Q is a closed convex set. We assume that the set Q is simple enough, so the gradient mapping can be computed explicitly. This assumption is valid for positive orthant, n dimensional box, simplex, Euclidean ball and some other sets.

Let us start from the gradient method:

Gradient method for simple sets

0. Choose $x_0 \in Q$. (2.2.18)

1. kth iteration ($k \geq 0$).

$$x_{k+1} = x_k - h\, g_Q(x_k; L).$$

The efficiency analysis of this scheme is very similar to that of the unconstrained version. Let us give an example of such a reasoning.

THEOREM 2.2.8 *Let $f \in \mathcal{S}^{1,1}_{\mu,L}(R^n)$. If in scheme (2.2.18) $h = \frac{1}{L}$, then*

$$\| x_k - x^* \|^2 \leq \left(1 - \frac{\mu}{L}\right)^k \| x_0 - x^* \|^2 .$$

Proof: Denote $r_k = \| x_k - x^* \|$, $g_Q = g_Q(x_k; L)$. Then, using inequality (2.2.17), we obtain

$$\begin{aligned} r_{k+1}^2 &= \| x_k - x^* - h g_Q \|^2 = r_k^2 - 2h\langle g_Q, x_k - x^*\rangle + h^2 \| g_Q \|^2 \\ &\leq (1 - h\mu) r_k^2 + h\left(h - \tfrac{1}{L}\right) \| g_G \|^2 = \left(1 - \tfrac{\mu}{L}\right) r_k^2. \end{aligned}$$

□

Note that for the step size $h = \frac{1}{L}$ we have

$$x_{k+1} = x_k - \tfrac{1}{L} g_Q(x_k; L) = x_Q(x_k; L).$$

Consider now the optimal schemes. We give only a sketch of justification since it is very similar to that of Section 2.2.1.

First of all, we define the estimate sequence. Assume that $x_0 \in Q$. Define

$$\begin{aligned} \phi_0(x) = {} & f(x_0) + \tfrac{\gamma_0}{2} \| x - x_0 \|^2, \\ \phi_{k+1}(x) = {} & (1 - \alpha_k)\phi_k(x) + \alpha_k[f(x_Q(y_k; L)) + \tfrac{1}{2L} \| g_Q(y_k; L) \|^2 \\ & + \langle g_Q(y_k; L), x - y_k\rangle + \tfrac{\mu}{2} \| x - y_k \|^2]. \end{aligned}$$

Note that the form of the recursive rule for $\phi_k(x)$ is changed. The reason is that now we use inequality (2.2.15) instead of (2.1.16). However, this modification does not change the analytical form of recursion and therefore it is possible to keep all convergence results of Section 2.2.1.

Similarly, it is easy to see that the estimate sequence $\{\phi_k(x)\}$ can be written as

$$\phi_k(x) = \phi_k^* + \tfrac{\gamma_k}{2} \parallel x - v_k \parallel^2,$$

with the following recursive rules for γ_k, v_k and ϕ_k^*:

$$\begin{aligned}
\gamma_{k+1} &= (1-\alpha_k)\gamma_k + \alpha_k\mu,\\
v_{k+1} &= \tfrac{1}{\gamma_{k+1}}[(1-\alpha_k)\gamma_k v_k + \alpha_k\mu y_k - \alpha_k g_Q(y_k;L)],\\
\phi_{k+1}^* &= (1-\alpha_k)\phi_k + \alpha_k f(x_Q(y_k;L))\\
&\quad + \left(\tfrac{\alpha_k}{2L} - \tfrac{\alpha_k^2}{2\gamma_{k+1}}\right) \parallel g_Q(y_k;L) \parallel^2\\
&\quad + \tfrac{\alpha_k(1-\alpha_k)\gamma_k}{\gamma_{k+1}} \left(\tfrac{\mu}{2} \parallel y_k - v_k \parallel^2 + \langle g_Q(y_k;L), v_k - y_k\rangle\right).
\end{aligned}$$

Further, assuming that $\phi_k^* \geq f(x_k)$ and using the inequality

$$\begin{aligned}
f(x_k) \;\geq\; & f(x_Q(y_k;L)) + \langle g_Q(y_k;L), x_k - y_k\rangle\\
& + \tfrac{1}{2L} \parallel g_Q(y_k;L) \parallel^2 + \tfrac{\mu}{2} \parallel x_k - y_k \parallel^2],
\end{aligned}$$

we come to the following lower bound:

$$\begin{aligned}
\phi_{k+1}^* \;\geq\; & (1-\alpha_k)f(x_k) + \alpha_k f(x_Q(y_k;L))\\
& + \left(\tfrac{\alpha_k}{2L} - \tfrac{\alpha_k^2}{2\gamma_{k+1}}\right) \parallel g_Q(y_k;L) \parallel^2\\
& + \tfrac{\alpha_k(1-\alpha_k)\gamma_k}{\gamma_{k+1}} \langle g_Q(y_k;L), v_k - y_k\rangle\\
\geq\; & f(x_Q(y_k;L)) + \left(\tfrac{1}{2L} - \tfrac{\alpha_k^2}{2\gamma_{k+1}}\right) \parallel g_Q(y_k;L) \parallel^2\\
& + (1-\alpha_k)\langle g_Q(y_k;L), \tfrac{\alpha_k\gamma_k}{\gamma_{k+1}}(v_k - y_k) + x_k - y_k\rangle.
\end{aligned}$$

Thus, again we can choose

$$x_{k+1} = x_Q(y_k; L),$$

$$L\alpha_k^2 = (1-\alpha_k)\gamma_k + \alpha_k \mu \equiv \gamma_{k+1},$$

$$y_k = \frac{1}{\gamma_k + \alpha_k \mu}(\alpha_k \gamma_k v_k + \gamma_{k+1} x_k).$$

Let us write down the corresponding variant of scheme (2.2.9).

Constant Step Scheme, II. Simple sets.

0. Choose $x_0 \in R^n$ and $\alpha_0 \in (0,1)$.
Set $y_0 = x_0$ and $q = \frac{\mu}{L}$.

1. kth iteration ($k \geq 0$).

a). Compute $f(y_k)$ and $f'(y_k)$. Set

$$x_{k+1} = x_Q(y_k; L). \tag{2.2.19}$$

b). Compute $\alpha_{k+1} \in (0,1)$ from equation

$$\alpha_{k+1}^2 = (1-\alpha_{k+1})\alpha_k^2 + q\alpha_{k+1},$$

and set $\beta_k = \frac{\alpha_k(1-\alpha_k)}{\alpha_k^2 + \alpha_{k+1}}$,

$$y_{k+1} = x_{k+1} + \beta_k(x_{k+1} - x_k).$$

Clearly, the rate of convergence of this method is given by Theorem 2.2.3. In this scheme only points $\{x_k\}$ are feasible for Q. The sequence $\{y_k\}$ is used for computing the gradient mapping and may be infeasible.

2.3 Minimization problem with smooth components

(Minimax problem: gradient mapping, gradient method, optimal methods; Problem with functional constraints; Methods for constrained minimization.)

2.3.1 Minimax problem

Very often the objective function of an optimization problem is composed by several components. For example, the reliability of a complex system usually is defined as a minimal reliability of its parts. A constrained minimization problem with functional constraints provides us with an example of interaction of several nonlinear functions, etc.

The simplest problem of that type is called the *minimax* problem. In this section we deal with the *smooth* minimax problem:

$$\min_{x \in Q} \left[f(x) = \max_{1 \le i \le m} f_i(x) \right], \tag{2.3.1}$$

where $f_i \in \mathcal{S}^{1,1}_{\mu,L}(R^n)$, $i = 1 \dots m$ and Q is a closed convex set. We call the function $f(x)$ *max-type* function composed by the *components* $f_i(x)$. We write $f \in \mathcal{S}^{1,1}_{\mu,L}(R^n)$ if all components of function f belong to that class.

Note that, in general, $f(x)$ is not differentiable. However, provided that all f_i are differentiable functions, we can introduce an object, which behaves exactly as a linear approximation of a smooth function.

DEFINITION 2.3.1 *Let f be a max-type function:*

$$f(x) = \max_{1 \le i \le m} f_i(x).$$

Function

$$f(\bar{x}; x) = \max_{1 \le i \le m} [f_i(\bar{x}) + \langle f_i'(\bar{x}), x - \bar{x} \rangle],$$

is called the linearization *of $f(x)$ at $\bar{x}$.*

Compare the following result with inequalities (2.1.16) and (2.1.6).

LEMMA 2.3.1 *For any $x \in R^n$ we have*

$$f(x) \ge f(\bar{x}; x) + \tfrac{\mu}{2} \| x - \bar{x} \|^2, \tag{2.3.2}$$

$$f(x) \le f(\bar{x}; x) + \tfrac{L}{2} \| x - \bar{x} \|^2 . \tag{2.3.3}$$

Proof: Indeed,

$$f_i(x) \ge f_i(\bar{x}) + \langle f_i'(\bar{x}), x - \bar{x} \rangle + \tfrac{\mu}{2} \| x - \bar{x} \|^2$$

(see (2.1.16)). Taking the maximum of this inequality in i, we get (2.3.2).

For (2.3.3) we use inequality

$$f_i(x) \le f_i(\bar{x}) + \langle f_i'(\bar{x}), x - \bar{x} \rangle + \tfrac{L}{2} \| x - \bar{x} \|^2$$

(see (2.1.6)). □

Let us write down the optimality conditions for problem (2.3.1) (compare with Theorem 2.2.5).

THEOREM 2.3.1 *A point $x^* \in Q$ is a solution to (2.3.1) if and only if for any $x \in Q$ we have*

$$f(x^*; x) \geq f(x^*; x^*) = f(x^*). \tag{2.3.4}$$

Proof: Indeed, if (2.3.4) is true, then

$$f(x) \geq f(x^*; x) \geq f(x^*; x^*) = f(x^*)$$

for all $x \in Q$.

Let x^* be a solution to (2.3.1). Assume that there exists $x \in Q$ such that $f(x^*; x) < f(x^*)$. Consider the functions

$$\phi_i(\alpha) = f_i(x^* + \alpha(x - x^*)), \quad i = 1 \dots m.$$

Note that for all i, $1 \leq i \leq m$, we have

$$f_i(x^*) + \langle f_i'(x^*), x - x^* \rangle < f(x^*) = \max_{1 \leq i \leq m} f_i(x^*).$$

Therefore either $\phi_i(0) \equiv f_i(x^*) < f(x^*)$, or

$$\phi_i(0) = f(x^*), \quad \phi_i'(0) = \langle f_i'(x^*), x - x^* \rangle < 0.$$

Therefore, for α small enough we have

$$f_i(x^* + \alpha(x - x^*)) = \phi_i(\alpha) < f(x^*)$$

for all i, $1 \leq i \leq m$. That is a contradiction. □

COROLLARY 2.3.1 *Let x^* be a minimum of a max-type function $f(x)$ on the set Q. If f belongs to $S^1_\mu(R^n)$, then*

$$f(x) \geq f(x^*) + \tfrac{\mu}{2} \| x - x^* \|^2$$

for all $x \in Q$.

Proof: Indeed, in view of (2.3.2) and Theorem 2.3.1, for any $x \in Q$ we have

$$\begin{aligned} f(x) &\geq f(x^*; x) + \tfrac{\mu}{2} \| x - x^* \|^2 \\ &\geq f(x^*; x^*) + \tfrac{\mu}{2} \| x - x^* \|^2 = f(x^*) + \tfrac{\mu}{2} \| x - x^* \|^2 . \end{aligned}$$

□

Finally, let us prove the existence theorem.

THEOREM 2.3.2 *Let max-type function $f(x)$ belong to $\mathcal{S}^1_\mu(R^n)$ with $\mu > 0$, and Q be a closed convex set. Then there exists a unique optimal solution x^* to the problem (2.3.1).*

Proof: Let $\bar{x} \in Q$. Consider the set $\bar{Q} = \{x \in Q \mid f(x) \leq f(\bar{x})\}$. Note that the problem (2.3.1) is equivalent to

$$\min\{f(x) \mid x \in \bar{Q}\}. \tag{2.3.5}$$

But $\bar{Q}$ is bounded: for any $x \in \bar{Q}$ we have

$$f(\bar{x}) \geq f_i(x) \geq f_i(\bar{x}) + \langle f_i'(\bar{x}), x - \bar{x}\rangle + \tfrac{\mu}{2} \parallel x - \bar{x} \parallel^2,$$

consequently,

$$\tfrac{\mu}{2} \parallel x - \bar{x} \parallel^2 \leq \parallel f'(\bar{x}) \parallel \cdot \parallel x - \bar{x} \parallel + f(\bar{x}) - f_i(\bar{x}).$$

Thus, the solution x^* of (2.3.5) (and of (2.3.1)) exists.

If x_1^* is another solution to (2.3.1), then

$$f(x^*) = f(x_1^*) \geq f(x^*; x_1^*) + \tfrac{\mu}{2} \parallel x_1^* - x^* \parallel^2 \geq f(x^*) + \tfrac{\mu}{2} \parallel x_1^* - x^* \parallel^2$$

(by (2.3.2)). Therefore $x_1^* = x^*$. □

2.3.2 Gradient mapping

In Section 2.2.3 we have introduced the gradient mapping, which replaces the gradient for a constrained minimization problem over a simple set. Since linearization of a max-type function behaves similarly to linearization of a smooth function, we can try to adapt the notion of gradient mapping to our particular situation.

Let us fix some $\gamma > 0$ and $\bar{x} \in R^n$. Consider a max-type function $f(x)$. Denote

$$f_\gamma(\bar{x}; x) = f(\bar{x}; x) + \tfrac{\gamma}{2} \parallel x - \bar{x} \parallel^2 .$$

The following definition is an extension of Definition 2.2.3.

DEFINITION 2.3.2 *Define*

$$\begin{aligned} f^*(\bar{x}; \gamma) &= \min_{x \in Q} f_\gamma(\bar{x}; x), \\ x_f(\bar{x}; \gamma) &= \arg\min_{x \in Q} f_\gamma(\bar{x}; x), \\ g_f(\bar{x}; \gamma) &= \gamma(\bar{x} - x_f(\bar{x}; \gamma)). \end{aligned}$$

We call $g_f(x;\gamma)$ gradient mapping *of max-type function* f *on* Q.

For $m = 1$ this definition is equivalent to Definition 2.2.3. Similarly, the point of *linearization* $\bar{x}$ does not necessarily belong to Q.

It is clear that $f_\gamma(\bar{x};x)$ is a max-type function composed by the components

$$f_i(\bar{x}) + \langle f_i'(\bar{x}), x - \bar{x}\rangle + \frac{\gamma}{2} \| x - \bar{x} \|^2 \in \mathcal{S}^{1,1}_{\gamma,\gamma}(R^n), \quad i = 0 \dots m.$$

Therefore the gradient mapping is well defined (Theorem 2.3.2).

Let us prove the main result of this section, which highlights the similarity between the properties of the gradient mapping and the properties of the gradient (compare with Theorem 2.2.7).

THEOREM 2.3.3 *Let* $f \in \mathcal{S}^{1,1}_{\mu,L}(R^n)$. *Then for all* $x \in Q$ *we have*

$$f(\bar{x};x) \geq f^*(\bar{x};\gamma) + \langle g_f(\bar{x};\gamma), x - \bar{x}\rangle + \frac{1}{2\gamma} \| g_f(\bar{x};\gamma) \|^2 . \qquad (2.3.6)$$

Proof: Denote $x_f = x_f(\bar{x};\gamma)$, $g_f = g_f(\bar{x};\gamma)$. It is clear that $f_\gamma(\bar{x};x) \in \mathcal{S}^{1,1}_{\gamma,\gamma}(R^n)$ and it is a max-type function. Therefore all results of the previous section can be applied also to f_γ.

Since $x_f = \arg\min\limits_{x\in Q} f_\gamma(\bar{x};x)$, in view of Corollary 2.3.1 and Theorem 2.3.1 we have

$$\begin{aligned} f(\bar{x};x) &= f_\gamma(\bar{x};x) - \frac{\gamma}{2} \| x - \bar{x} \|^2 \\ &\geq f_\gamma(\bar{x};x_f) + \frac{\gamma}{2}(\| x - x_f \|^2 - \| x - \bar{x} \|^2) \\ &\geq f^*(\bar{x};\gamma) + \frac{\gamma}{2}\langle \bar{x} - x_f, 2x - x_f - \bar{x}\rangle \\ &= f^*(\bar{x};\gamma) + \frac{\gamma}{2}\langle \bar{x} - x_f, 2(x - \bar{x}) + \bar{x} - x_f\rangle \\ &= f^*(\bar{x};\gamma) + \langle g_f, x - \bar{x}\rangle + \frac{1}{2\gamma} \| g_f \|^2 . \end{aligned}$$

□

In what follows we often refer to the following corollary to Theorem 2.3.3.

COROLLARY 2.3.2 *Let* $f \in \mathcal{S}^{1,1}_{\mu,L}(R^n)$ *and* $\gamma \geq L$. *Then:*

1. For any $x \in Q$ *and* $\bar{x} \in R^n$ *we have*

$$\begin{aligned} f(x) \geq\ & f(x_f(\bar{x};\gamma)) + \langle g_f(\bar{x};\gamma), x - \bar{x}\rangle \\ & + \frac{1}{2\gamma} \| g_f(\bar{x};\gamma) \|^2 + \frac{\mu}{2} \| x - \bar{x} \|^2 . \end{aligned} \qquad (2.3.7)$$

2. If $\bar{x} \in Q$, then

$$f(x_f(\bar{x};\gamma)) \leq f(\bar{x}) - \tfrac{1}{2\gamma} \parallel g_f(\bar{x};\gamma) \parallel^2, \qquad (2.3.8)$$

3. For any $\bar{x} \in R^n$ we have

$$\langle g_f(\bar{x};\gamma), \bar{x} - x^* \rangle \geq \tfrac{1}{2\gamma} \parallel g_f(\bar{x};\gamma) \parallel^2 + \tfrac{\mu}{2} \parallel x^* - \bar{x} \parallel^2 . \qquad (2.3.9)$$

Proof: Assumption $\gamma \geq L$ implies that $f^*(\bar{x};\gamma) \geq f(x_f(\bar{x};\gamma))$. Therefore (2.3.7) follows from (2.3.6) since

$$f(x) \geq f(\bar{x};x) + \tfrac{\mu}{2} \parallel x - \bar{x} \parallel^2$$

for all $x \in R^n$ (see Lemma 2.3.1).

Using (2.3.7) with $x = \bar{x}$, we get (2.3.8), and using (2.3.7) with $x = x^*$, we get (2.3.9) since $f(x_f(\bar{x};\gamma)) - f(x^*) \geq 0$. □

Finally, let us estimate the variation of $f^*(\bar{x};\gamma)$ as a function of γ.

LEMMA 2.3.2 *For any γ_1, $\gamma_2 > 0$ and $\bar{x} \in R^n$ we have*

$$f^*(\bar{x};\gamma_2) \geq f^*(\bar{x};\gamma_1) + \tfrac{\gamma_2 - \gamma_1}{2\gamma_1\gamma_2} \parallel g_f(\bar{x};\gamma_1) \parallel^2 .$$

Proof: Denote $x_i = x_f(\bar{x};\gamma_i)$, $g_i = g_f(\bar{x};\gamma_i)$, $i = 1, 2$. In view of (2.3.6), we have

$$\begin{aligned} f(\bar{x};x) + \tfrac{\gamma_2}{2} \parallel x - \bar{x} \parallel^2 \;\; &\geq \;\; f^*(\bar{x};\gamma_1) + \langle g_1, x - \bar{x} \rangle \\ &\quad + \tfrac{1}{2\gamma_1} \parallel g_1 \parallel^2 + \tfrac{\gamma_2}{2} \parallel x - \bar{x} \parallel^2 \end{aligned} \qquad (2.3.10)$$

for all $x \in Q$. In particular, for $x = x_2$ we obtain

$$\begin{aligned} f^*(\bar{x};\gamma_2) \;\; &= \;\; f(\bar{x};x_2) + \tfrac{\gamma_2}{2} \parallel x_2 - \bar{x} \parallel^2 \\ &\geq \;\; f^*(\bar{x};\gamma_1) + \langle g_1, x_2 - \bar{x} \rangle + \tfrac{1}{2\gamma_1} \parallel g_1 \parallel^2 + \tfrac{\gamma_2}{2} \parallel x_2 - \bar{x} \parallel^2 \\ &= \;\; f^*(\bar{x};\gamma_1) + \tfrac{1}{2\gamma_1} \parallel g_1 \parallel^2 - \tfrac{1}{\gamma_2} \langle g_1, g_2 \rangle + \tfrac{1}{2\gamma_2} \parallel g_2 \parallel^2 \\ &\geq \;\; f^*(\bar{x};\gamma_1) + \tfrac{1}{2\gamma_1} \parallel g_1 \parallel^2 - \tfrac{1}{2\gamma_2} \parallel g_1 \parallel^2 . \end{aligned}$$

□

2.3.3 Minimization methods for minimax problem

As usual, we start a presentation of numerical methods for problem (2.3.1) from a "gradient" method with constant step:

Gradient method for minimax problem
0. Choose $x_0 \in Q$ and $h > 0$.
1. kth iteration ($k \geq 0$). $$x_{k+1} = x_k - hg_f(x_k; L).$$

(2.3.11)

THEOREM 2.3.4 *Let $f \in S^{1,1}_{\mu,L}(R^n)$. If in (2.3.11) we choose $h \leq \frac{1}{L}$, then*

$$\| x_k - x^* \|^2 \leq (1 - \mu h)^k \| x_0 - x^* \|^2 .$$

Proof: Denote $r_k = \| x_k - x^* \|$, $g = g_f(x_k; L)$. Then, in view of (2.3.9) we have

$$\begin{aligned} r^2_{k+1} &= \| x_k - x^* - hg_Q \|^2 = r_k^2 - 2h\langle g, x_k - x^* \rangle + h^2 \| g \|^2 \\ &\leq (1 - h\mu)r_k^2 + h\left(h - \tfrac{1}{L}\right) \| g \|^2 \leq (1 - \mu h) r_k^2. \end{aligned}$$

□

Note that with $h = \frac{1}{L}$ we have

$$x_{k+1} = x_k - \tfrac{1}{L} g_f(x_k; L) = x_f(x_k; L).$$

For this step size, the rate of convergence of scheme (2.3.11) is as follows:

$$\| x_k - x^* \|^2 \leq \left(1 - \tfrac{\mu}{L}\right)^k \| x_0 - x^* \|^2 .$$

Comparing this result with Theorem 2.2.8, we see that for minimax problem the gradient method has the same rate of convergence, as it has in the smooth case.

Let us check, what the situation is with the optimal methods. Recall, that in order to develop an optimal method, we need to introduce an estimate sequence with some recursive updating rules. Formally, the minimax problem differs from the unconstrained minimization problem only by the form of lower approximation of the objective function. In the case of unconstrained minimization, inequality (2.1.16) was used for updating the estimate sequence. Now it must be replaced by inequality (2.3.7).

Let us introduce an estimate sequence for problem (2.3.1). Let us fix some $x_0 \in Q$ and $\gamma_0 > 0$. Consider the sequences $\{y_k\} \subset R^n$ and $\{\alpha_k\} \subset (0,1)$. Define

$$\begin{aligned}
\phi_0(x) &= f(x_0) + \tfrac{\gamma_0}{2} \parallel x - x_0 \parallel^2, \\
\phi_{k+1}(x) &= (1-\alpha_k)\phi_k(x) \\
&\quad + \alpha_k[\boxed{\ f(x_f(y_k;L)) + \tfrac{1}{2L} \parallel g_f(y_k;L) \parallel^2} \\
&\quad + \langle g_f(y_k;L), x - y_k\rangle + \tfrac{\mu}{2} \parallel x - y_k \parallel^2].
\end{aligned}$$

Comparing these relations with (2.2.3), we can find the difference only in the constant term (it is in the frame). In (2.2.3) this place was taken by $f(y_k)$. This difference leads to a trivial modification in the results of Lemma 2.2.3: All inclusions of $f(y_k)$ must be formally replaced by the expression in the frame, and $f'(y_k)$ must be replaced by $g_f(y_k;L)$. Thus, we come to the following lemma.

LEMMA 2.3.3 *For all $k \geq 0$ we have*

$$\phi_k(x) \equiv \phi_k^* + \tfrac{\gamma_k}{2} \parallel x - v_k \parallel^2,$$

where the sequences $\{\gamma_k\}$, $\{v_k\}$ and $\{\phi_k^*\}$ are defined as follows: $v_0 = x_0$, $\phi_0^* = f(x_0)$ and

$$\begin{aligned}
\gamma_{k+1} &= (1-\alpha_k)\gamma_k + \alpha_k\mu, \\
v_{k+1} &= \tfrac{1}{\gamma_{k+1}}[(1-\alpha_k)\gamma_k v_k + \alpha_k \mu y_k - \alpha_k g_f(y_k;L)], \\
\phi_{k+1}^* &= (1-\alpha_k)\phi_k + \alpha_k(f(x_f(y_k;L)) + \tfrac{1}{2L} \parallel g_f(y_k;L) \parallel^2) \\
&\quad + \tfrac{\alpha_k^2}{2\gamma_{k+1}} \parallel g_f(y_k;L) \parallel^2 \\
&\quad + \tfrac{\alpha_k(1-\alpha_k)\gamma_k}{\gamma_{k+1}} \left(\tfrac{\mu}{2} \parallel y_k - v_k \parallel^2 + \langle g_f(y_k;L), v_k - y_k\rangle\right).
\end{aligned}$$

□

Now we can proceed exactly as in Section 2.2. Assume that $\phi_k^* \geq f(x_k)$. Inequality (2.3.7) with $x = x_k$ and $\bar{x} = y_k$ becomes

$$\begin{aligned}
f(x_k) &\geq f(x_f(y_k;L)) + \langle g_f(y_k;L), x_k - y_k\rangle \\
&\quad + \tfrac{1}{2L} \parallel g_f(y_k;L) \parallel^2 + \tfrac{\mu}{2} \parallel x_k - y_k \parallel^2 .
\end{aligned}$$

Hence,

$$\phi_{k+1}^* \geq (1-\alpha_k)f(x_k) + \alpha_k f(x_f(y_k;L))$$

$$+\left(\frac{\alpha_k}{2L} - \frac{\alpha_k^2}{2\gamma_{k+1}}\right) \| g_f(y_k;L) \|^2 + \frac{\alpha_k(1-\alpha_k)\gamma_k}{\gamma_{k+1}} \langle g_f(y_k;L), v_k - y_k \rangle$$

$$\geq f(x_f(y_k;L)) + \left(\frac{1}{2L} - \frac{\alpha_k^2}{2\gamma_{k+1}}\right) \| g_f(y_k;L) \|^2$$

$$+(1-\alpha_k)\langle g_f(y_k;L), \frac{\alpha_k\gamma_k}{\gamma_{k+1}}(v_k - y_k) + x_k - y_k \rangle.$$

Thus, again we can choose

$$\begin{aligned} x_{k+1} &= x_f(y_k;L), \\ L\alpha_k^2 &= (1-\alpha_k)\gamma_k + \alpha_k\mu \equiv \gamma_{k+1}, \\ y_k &= \frac{1}{\gamma_k+\alpha_k\mu}(\alpha_k\gamma_k v_k + \gamma_{k+1}x_k). \end{aligned}$$

Let us write down the resulting scheme in the form of (2.2.9), with eliminated sequences $\{v_k\}$ and $\{\gamma_k\}$.

Constant Step Scheme, II. Minimax.

0. Choose $x_0 \in R^n$ and $\alpha_0 \in (0,1)$.
Set $y_0 = x_0$ and $q = \frac{\mu}{L}$.

1. kth iteration ($k \geq 0$).

a). Compute $\{f_i(y_k)\}$ and $\{f_i'(y_k)\}$. Set

$$x_{k+1} = x_f(y_k;L).$$

b). Compute $\alpha_{k+1} \in (0,1)$ from equation

$$\alpha_{k+1}^2 = (1-\alpha_{k+1})\alpha_k^2 + q\alpha_{k+1},$$

and set $\beta_k = \frac{\alpha_k(1-\alpha_k)}{\alpha_k^2+\alpha_{k+1}}$,

$$y_{k+1} = x_{k+1} + \beta_k(x_{k+1} - x_k).$$

(2.3.12)

The convergence analysis of this scheme is completely identical to that of scheme (2.2.9). Let us just give the result.

THEOREM 2.3.5 *Let the max-type function f belong to $\mathcal{S}^{1,1}_{\mu,L}(R^n)$. If in (2.3.12) we take $\alpha_0 \geq \sqrt{\frac{\mu}{L}}$, then*

$$f(x_k) - f^* \leq \min\left\{\left(1-\sqrt{\frac{\mu}{L}}\right)^k, \frac{4L}{(2\sqrt{L}+k\sqrt{\gamma_0})^2}\right\}$$

$$\times \left[f(x_0) - f^* + \frac{\gamma_0}{2} \parallel x_0 - x^* \parallel^2\right],$$

where $\gamma_0 = \frac{\alpha_0(\alpha_0 L - \mu)}{1-\alpha_0}$.

□

Note that the scheme (2.3.12) works for all $\mu \geq 0$. Let us write down the method for solving (2.3.1) with strictly convex components.

Scheme for $f \in \mathcal{S}^{1,1}_{\mu,L}(R^n)$
0. Choose $x_0 \in Q$. Set $y_0 = x_0$, $\beta = \frac{\sqrt{L}-\sqrt{\mu}}{\sqrt{L}+\sqrt{\mu}}$.
1. kth iteration ($k \geq 0$). Compute $\{f_i(y_k)\}$ and $\{f_i'(y_k)\}$. Set $x_{k+1} = x_f(y_k; L), \quad y_{k+1} = x_{k+1} + \beta(x_{k+1} - x_k)$.

(2.3.13)

THEOREM 2.3.6 *For scheme (2.3.13) we have*

$$f(x_k) - f^* \leq 2\left(1-\sqrt{\frac{\mu}{L}}\right)^k (f(x_0) - f^*). \tag{2.3.14}$$

Proof: Scheme (2.3.13) is a variant of (2.3.12) with $\alpha_0 = \sqrt{\frac{\mu}{L}}$. Under this choice, $\gamma_0 = \mu$ and we get (2.3.14) from Theorem 2.3.5 since, in view of Corollary 2.3.1, $\frac{\mu}{2} \parallel x_0 - x^* \parallel^2 \leq f(x_0) - f^*$. □

To conclude this section, let us look at an auxiliary problem, which we need to solve in order to compute the gradient mapping of the minimax problem. Recall, that this problem is as follows:

$$\min_{x\in Q}\left\{\max_{1\le i\le m}[f_i(x_0)+\langle f_i'(x_0),x-x_0\rangle]+\tfrac{\gamma}{2}\parallel x-x_0\parallel^2\right\}.$$

Introducing the additional variables $t\in R^m$, we can rewrite this problem in the following way:

$$\begin{aligned}&\min\ \{\sum_{i=1}^{m} t^{(i)}+\tfrac{\gamma}{2}\parallel x-x_0\parallel^2\}\\ \text{s. t.}\quad & f_i(x_0)+\langle f_i'(x_0),x-x_0\rangle\le t^{(i)},\ i=1\dots m,\\ & x\in Q,\ t\in R^m,\end{aligned}\tag{2.3.15}$$

Note that if Q is a polytope, then the problem (2.3.15) is a quadratic optimization problem. This problem can be solved by some special finite methods (simplex-type algorithms). It can be also solved by interior point methods. In the latter case, we can treat much more complicated nonlinear structure of the set Q.

2.3.4 Optimization with functional constraints

Let us show that methods described in the previous section can be used for solving a constrained minimization problem with smooth functional constraints. Recall, that the analytical form of such a problem is as follows:

$$\begin{aligned}&\min\ f_0(x),\\ \text{s.t.}\quad & f_i(x)\le 0,\ i=1\dots m,\\ & x\in Q,\end{aligned}\tag{2.3.16}$$

where the functions f_i are convex and smooth and Q is a closed convex set. In this section we assume that $f_i\in \mathcal{S}^{1,1}_{\mu,L}(R^n)$, $i=0\dots m$, with some $\mu>0$.

The relation between the problem (2.3.16) and minimax problems is established by some special function of one variable. Consider the *parametric* max-type function

$$f(t;x)=\max\{f_0(x)-t;f_i(x),i=1\dots m\},\quad t\in R^1,\ x\in Q.$$

Let us introduce the function

$$f^*(t)=\min_{x\in Q} f(t;x).\tag{2.3.17}$$

Note that the components of max-type function $f(t;\cdot)$ are strongly convex in x. Therefore, for any $t \in R^1$ the solution of problem (2.3.17), $x^*(t)$, exists and is unique in view of Theorem 2.3.2.

We will try to get close to the solution of (2.3.16) using a process based on *approximate values* of function $f^*(t)$. This approach can be seen as a variant of *sequential quadratic optimization.* It can be applied also to nonconvex problems.

Let us establish some properties of function $f^*(t)$.

LEMMA 2.3.4 *Let t^* be an optimal value of problem (2.3.16). Then*

$$\begin{aligned} f^*(t) &\leq 0 \quad \textit{for all } t \geq t^*, \\ f^*(t) &> 0 \quad \textit{for all } t < t^*. \end{aligned}$$

Proof: Let x^* be a solution to (2.3.16). If $t \geq t^*$, then

$$\begin{aligned} f^*(t) &\leq f(t;x^*) = \max\{f_0(x^*) - t; f_i(x^*)\} \\ &\leq \max\{t^* - t; f_i(x^*)\} \leq 0. \end{aligned}$$

Suppose that $t < t^*$ and $f^*(t) \leq 0$. Then there exists $y \in Q$ such that

$$f_0(y) \leq t < t^*, \quad f_i(y) \leq 0,\ i = 1 \dots m.$$

Thus, t^* cannot be an optimal value of (2.3.16). □

Thus, the smallest root of function $f^*(t)$ corresponds to the optimal value of the problem (2.3.16). Note also, that using the methods of the previous section, we can compute an approximate value of function $f^*(t)$. Hence, our goal is to form a process of finding the root, based on that information. However, for that we need to know more properties of function $f^*(t)$.

LEMMA 2.3.5 *For any $\Delta \geq 0$ we have*

$$f^*(t) - \Delta \leq f^*(t + \Delta) \leq f^*(t).$$

Proof: Indeed,

$$\begin{aligned} f^*(t+\Delta) &= \min_{x\in Q} \max_{1\leq i\leq m} \{f_0(x) - t - \Delta; f_i(x)\} \\ &\leq \min_{x\in Q} \max_{1\leq i\leq m} \{f_0(x) - t; f_i(x)\} = f^*(t), \\ f^*(t+\Delta) &= \min_{x\in Q} \max_{1\leq i\leq m} \{f_0(x) - t; f_i(x) + \Delta\} - \Delta \\ &\geq \min_{x\in Q} \max_{1\leq i\leq m} \{f_0(x) - t; f_i(x)\} - \Delta = f^*(t) - \Delta. \end{aligned}$$

□

In other words, function $f^*(t)$ decreases in t and it is Lipschitz continuous with constant equal to 1.

LEMMA 2.3.6 *For any $t_1 < t_2$ and $\Delta \geq 0$ we have*

$$f^*(t_1 - \Delta) \geq f^*(t_1) + \Delta \tfrac{f^*(t_1) - f^*(t_2)}{t_2 - t_1}. \quad (2.3.18)$$

Proof: Denote $t_0 = t_1 - \Delta$, $\alpha = \frac{\Delta}{t_2 - t_0} \equiv \frac{\Delta}{t_2 - t_1 + \Delta} \in [0, 1]$. Then $t_1 = (1-\alpha)t_0 + \alpha t_2$ and (2.3.18) can be written as

$$f^*(t_1) \leq (1-\alpha) f^*(t_0) + \alpha f^*(t_2). \quad (2.3.19)$$

Let $x_\alpha = (1-\alpha)x^*(t_0) + \alpha x^*(t_2)$. We have

$$\begin{aligned} f^*(t_1) &\leq \max_{1\leq i\leq m} \{f_0(x_\alpha) - t_1; f_i(x_\alpha)\} \\ &\leq \max_{1\leq i\leq m} \{(1-\alpha)(f_0(x^*(t_0)) - t_0) + \alpha(f_0(x^*(t_2)) - t_2); \\ &\qquad (1-\alpha) f_i(x^*(t_0)) + \alpha f_i(x^*(t_2))\} \\ &\leq (1-\alpha) \max_{1\leq i\leq m} \{f_0(x^*(t_0)) - t_0; f_i(x^*(t_0))\} \\ &\qquad + \alpha \max_{1\leq i\leq m} \{f_0(x^*(t_2)) - t_2; f_i(x^*(t_2))\} \\ &= (1-\alpha) f^*(t_0) + \alpha f^*(t_2), \end{aligned}$$

and we get (2.3.18). □

Note that Lemmas 2.3.5 and 2.3.6 are valid for *any* parametric max-type functions, not necessarily formed by functional components of problem (2.3.16).

Let us study now the properties of gradient mapping for a parametric max-type function. To do that, let us introduce first a *linearization* of a parametric max-type function $f(t;x)$:

$$f(t;\bar{x};x) = \max_{1\leq i\leq m}\{f_0(\bar{x}) + \langle f_0'(\bar{x}), x-\bar{x}\rangle - t; f_i(\bar{x}) + \langle f_i'(\bar{x}), x-\bar{x}\rangle\}.$$

Now we can introduce a gradient mapping in a standard way. Let us fix some $\gamma > 0$. Denote

$$\begin{aligned} f_\gamma(t;\bar{x};x) &= f(t;\bar{x};x) + \tfrac{\gamma}{2}\parallel x-\bar{x}\parallel^2, \\ f^*(t;\bar{x};\gamma) &= \min_{x\in Q} f_\gamma(t;\bar{x};x) \\ x_f(t;\bar{x};\gamma) &= \arg\min_{x\in Q} f_\gamma(t;\bar{x};x) \\ g_f(t;\bar{x};\gamma) &= \gamma(\bar{x} - x_f(t;\bar{x};\gamma)). \end{aligned}$$

We call $g_f(t;\bar{x};\gamma)$ the *constrained gradient mapping* of problem (2.3.16). As usual, the point of linearization $\bar{x}$ is not necessarily feasible for Q.

Note that function $f_\gamma(t;\bar{x};x)$ itself is a max-type function composed by the components

$$f_0(\bar{x}) + \langle f_0'(\bar{x}), x-\bar{x}\rangle - t + \tfrac{\gamma}{2}\parallel x-\bar{x}\parallel^2,$$

$$f_i(\bar{x}) + \langle f_i'(\bar{x}), x-\bar{x}\rangle - t + \tfrac{\gamma}{2}\parallel x-\bar{x}\parallel^2,\ i = 1\ldots m.$$

Moreover, $f_\gamma(t;\bar{x};x) \in \mathcal{S}^{1,1}_{\gamma,\gamma}(R^n)$. Therefore, for any $t \in R^1$ the constrained gradient mapping is well defined in view of Theorem 2.3.2.

Since $f(t;x) \in \mathcal{S}^{1,1}_{\mu,L}(R^n)$, we have

$$f_\mu(t;\bar{x};x) \leq f(t;x) \leq f_L(t;\bar{x};x)$$

for all $x \in R^n$. Therefore $f^*(t;\bar{x};\mu) \leq f^*(t) \leq f^*(t;\bar{x};L)$. Moreover, using Lemma 2.3.6, we obtain the following result:

For any $\bar{x} \in R^n$, $\gamma > 0$, $\Delta \geq 0$ and $t_1 < t_2$ and we have

$$f^*(t_1-\Delta;\bar{x};\gamma) \geq f^*(t_1;\bar{x};\gamma) + \tfrac{\Delta}{t_2-t_1}(f^*(t_1;\bar{x};\gamma) - f^*(t_2;\bar{x};\gamma)). \tag{2.3.20}$$

There are two values of γ, which are important for us. These are $\gamma = L$ and $\gamma = \mu$. Applying Lemma 2.3.2 to max-type function $f_\gamma(t;\bar{x};x)$ with

$\gamma_1 = L$ and $\gamma_2 = \mu$, we obtain the following inequality:

$$f^*(t; \bar{x}; \mu) \geq f^*(t; \bar{x}; L) - \frac{L-\mu}{2\mu L} \parallel g_f(t; \bar{x}; L) \parallel^2 . \qquad (2.3.21)$$

Since we are interested in finding a root of the function $f^*(t)$, let us describe the behavior of the roots of function $f^*(t; \bar{x}; \gamma)$, which can be seen as an approximation of $f^*(t)$.

Denote

$$t^*(\bar{x}, t) = \text{root}_t(f^*(t; \bar{x}; \mu))$$

(notation $\text{root}_t(\cdot)$ means the root in t of function $(\cdot)$).

LEMMA 2.3.7 *Let $\bar{x} \in R^n$ and $\bar{t} < t^*$ be such that*

$$f^*(\bar{t}; \bar{x}; \mu) \geq (1-\kappa) f^*(\bar{t}; \bar{x}; L)$$

for some $\kappa \in (0,1)$. Then $\bar{t} < t^*(\bar{x}, \bar{t}) \leq t^*$. Moreover, for any $t < \bar{t}$ and $x \in R^n$ we have

$$f^*(t; x; L) \geq 2(1-\kappa) f^*(\bar{t}; \bar{x}; L) \sqrt{\frac{\bar{t}-t}{t^*(\bar{x}, \bar{t}) - \bar{t}}}.$$

Proof: Since $\bar{t} < t^*$, we have

$$0 < f^*(\bar{t}) \leq f^*(\bar{t}; \bar{x}; L) \leq \frac{1}{1-\kappa} f^*(\bar{t}; \bar{x}; \mu).$$

Thus, $f^*(\bar{t}; \bar{x}; \mu) > 0$ and, since $f^*(t; \bar{x}; \mu)$ decreases in t, we get

$$t^*(\bar{x}, \bar{t}) > \bar{t}.$$

Denote $\Delta = \bar{t} - t$. Then, in view of (2.3.20), we have

$$\begin{aligned} f^*(t; x; L) &\geq f^*(t) \geq f^*(t; \bar{x}; \mu) \geq f^*(\bar{t}; \bar{x}; \mu) + \frac{\Delta}{t^*(\bar{x}, \bar{t}) - \bar{t}} f^*(\bar{t}; \bar{x}; \mu) \\ &\geq (1-\kappa)\left(1 + \frac{\Delta}{t^*(\bar{x}, \bar{t}) - \bar{t}}\right) f^*(\bar{t}; \bar{x}; L) \\ &\geq 2(1-\kappa) f^*(\bar{t}; \bar{x}; L) \sqrt{\frac{\Delta}{t^*(\bar{x}, \bar{t}) - \bar{t}}}. \end{aligned}$$

□

2.3.5 Method for constrained minimization

Now we are ready to analyze the following process.

Constrained minimization scheme

0. Choose $x_0 \in Q$, $\kappa \in (0, \frac{1}{2})$, $t_0 < t^*$ and accuracy $\epsilon > 0$.

1. kth iteration ($k \geq 0$).

a). Generate sequence $\{x_{k,j}\}$ by method (2.3.13) as applied to $f(t_k; x)$ with starting point $x_{k,0} = x_k$. If

$$f^*(t_k; x_{k,j}; \mu) \geq (1 - \kappa) f^*(t_k; x_{k,j}; L) \tag{2.3.22}$$

then stop the internal process and set $j(k) = j$,

$$j^*(k) = \arg \min_{0 \leq j \leq j(k)} f^*(t_k; x_{k,j}; L),$$

$$x_{k+1} = x_f(t_k; x_{k,j^*(k)}; L).$$

Global stop, if at some iteration of the internal scheme we have $f^*(t_k; x_{k,j}; L) \leq \epsilon$.

b). Set $t_{k+1} = t^*(x_{k,j(k)}, t_k)$.

This is the first time in our course we meet a two-level process. Clearly, its analysis is rather complicated. Firstly, we need to estimate the rate of convergence of the upper-level process in (2.3.22) (it is called a *master process*). Secondly, we need to estimate the total complexity of the internal processes in Step 1a). Since we are interested in the analytical complexity of this method, the arithmetical cost of computation of $t^*(x, t)$ and $f^*(t; x, \gamma)$ is not important for us now.

Let us describe convergence of the master process.

LEMMA 2.3.8

$$f^*(t_k; x_{k+1}; L) \leq \frac{t_0 - t^*}{1 - \kappa} \left[\frac{1}{2(1-\kappa)} \right]^k.$$

Proof: Denote $\beta = \frac{1}{2(1-\kappa)}$ (< 1) and

$$\delta_k = \frac{f^*(t_k; x_{k,j(k)}; L)}{\sqrt{t_{k+1} - t_k}}.$$

Since $t_{k+1} = t^*(x_{k,j(k)}, t_k)$, in view of Lemma 2.3.7 for $k \geq 1$ we have

$$2(1-\kappa)\frac{f^*(t_k;x_{k,j(k)};L)}{\sqrt{t_{k+1}-t_k}} \leq \frac{f^*(t_{k-1};x_{k-1,j(k-1)};L))}{\sqrt{t_k - t_{k-1}}}.$$

Thus, $\delta_k \leq \beta\delta_{k-1}$ and we obtain

$$\begin{aligned} f^*(t_k; x_{k,j(k)}; L) &= \delta_k\sqrt{t_{k+1}-t_k} \leq \beta^k\delta_0\sqrt{t_{k+1}-t_k} \\ &= \beta^k f^*(t_0; x_{0,j(0)}; L)\sqrt{\frac{t_{k+1}-t_k}{t_1-t_0}}. \end{aligned}$$

Further, in view of Lemma 2.3.5, we have $t_1 - t_0 \geq f^*(t_0; x_{0,j(0)}; \mu)$. Hence,

$$\begin{aligned} f^*(t_k; x_{k,j(k)}; L) &\leq \beta^k f^*(t_0; x_{0,j(0)}; L)\sqrt{\frac{t_{k+1}-t_k}{f^*(t_0;x_{0,j(0)};\mu)}} \\ &\leq \frac{\beta^k}{1-\kappa}\sqrt{f^*(t_0; x_{0,j(0)}; \mu)(t_{k+1}-t_k)} \\ &\leq \frac{\beta^k}{1-\kappa}\sqrt{f^*(t_0)(t_0-t^*)}. \end{aligned}$$

It remains to note that $f^*(t_0) \leq t_0 - t^*$ (Lemma 2.3.5) and

$$f^*(t_k; x_{k+1}; L) \equiv f^*(t_k; x_{k,j^*(k)}; L) \leq f^*(t_k; x_{k,j(k)}; L).$$

□

The above result provides us with an estimate for the number of upper-level iterations, which are necessary to find an ϵ-solution of problem (2.3.16). Indeed, let $f^*(t_k; x_{k,j}; L) \leq \epsilon$. Then for $x_* = x_f(t_k; x_{k,j}; L)$ we have

$$f(t_k; x_*) = \max_{1\leq i\leq m}\{f_0(x_*) - t_k; f_i(x_*)\} \leq f^*(t_k; x_{k,j}; L) \leq \epsilon.$$

Since $t_k \leq t^*$, we conclude that

$$\begin{aligned} f_0(x_*) &\leq t^* + \epsilon, \\ f_i(x_*) &\leq \epsilon, \; i = 1\ldots m. \end{aligned} \tag{2.3.23}$$

In view of Lemma 2.3.8, we can get (2.3.23) at most in

$$N(\epsilon) = \frac{1}{\ln[2(1-\kappa)]}\ln\frac{t_0-t^*}{(1-\kappa)\epsilon} \tag{2.3.24}$$

full iterations of the master process (the last iteration of the process, in general, is not full since it is terminated by the Global stop rule). Note that in this estimate κ is an absolute constant (for example, $\kappa = \frac{1}{4}$).

Let us analyze the complexity of the internal process. Let the sequence $\{x_{k,j}\}$ be generated by (2.3.13) with the starting point $x_{k,0} = x_k$. In view of Theorem 2.3.6, we have

$$\begin{aligned} f(t_k; x_{k,j}) - f^*(t_k) &\le 2\left(1 - \sqrt{\tfrac{\mu}{L}}\right)^j (f(t_k; x_k) - f^*(t_k)) \\ &\le 2e^{-\sigma \cdot j}(f(t_k; x_k) - f^*(t_k)) \le 2e^{-\sigma \cdot j} f(t_k; x_k), \end{aligned}$$

where $\sigma = \sqrt{\frac{\mu}{L}}$.

Denote by N the number of full iterations of the process (2.3.22) ($N \le N(\epsilon)$). Thus, $j(k)$ is defined for all k, $0 \le k \le N$. Note that $t_k = t^*(x_{k-1,j(k-1)}, t_{k-1}) > t_{k-1}$. Therefore

$$f(t_k; x_k) \le f(t_{k-1}; x_k) \le f^*(t_{k-1}; x_{k-1,j^*(k-1)}, L).$$

Denote

$$\Delta_k = f^*(t_{k-1}; x_{k-1,j^*(k-1)}, L), \quad k \ge 1, \quad \Delta_0 = f(t_0; x_0).$$

Then, for all $k \ge 0$ we have

$$f(t_k; x_k) - f^*(t_k) \le \Delta_k.$$

LEMMA 2.3.9 *For all k, $0 \le k \le N$, the internal process works no longer as the following condition is satisfied:*

$$f(t_k; x_{k,j}) - f^*(t_k) \le \tfrac{\mu\kappa}{L-\mu} \cdot f^*(t_k; x_{k,j}; L). \qquad (2.3.25)$$

Proof: Assume that (2.3.25) is satisfied. Then, in view of (2.3.8), we have

$$\begin{aligned} \tfrac{1}{2L} \| g_f(t_k; x_{k,j}; L \|^2 &\le f(t_k; x_{k,j}) - f(t_k; x_f(t_k; x_{k,j}; L)) \\ &\le f(t_k; x_{k,j}) - f^*(t_k). \end{aligned}$$

Therefore, using (2.3.21), we obtain

$$\begin{aligned} f^*(t_k; x_{k,j}; \mu) &\ge f^*(t_k; x_{k,j}; L) - \tfrac{L-\mu}{2\mu L} \| g_f(t_k; x_{k,j}; L \|^2 \\ &\ge f^*(t_k; x_{k,j}; L) - \tfrac{L-\mu}{\mu}(f(t_k; x_{k,j}) - f^*(t_k)) \\ &\ge (1 - \kappa) f^*(t_k; x_{k,j}; L). \end{aligned}$$

And that is the termination criterion of the internal process in Step 1a) in (2.3.22). □

The above result, combined with the estimate of the rate of convergence for the internal process, provide us with the total complexity estimate of the constrained minimization scheme.

LEMMA 2.3.10 *For all k, $0 \le k \le N$, we have*

$$j(k) \le 1 + \sqrt{\tfrac{L}{\mu}} \cdot \ln \tfrac{2(L-\mu)\Delta_k}{\kappa\mu\Delta_{k+1}}.$$

Proof: Assume that

$$j(k) - 1 > \tfrac{1}{\sigma} \ln \tfrac{2(L-\mu)\Delta_k}{\kappa\mu\Delta_{k+1}},$$

where $\sigma = \sqrt{\frac{\mu}{L}}$. Recall that $\Delta_{k+1} = \min\limits_{0 \le j \le j(k)} f^*(t_k; x_{k,j}; L)$. Note that the stopping criterion of the internal process did not work for $j = j(k) - 1$. Therefore, in view of Lemma 2.3.9, we have

$$f^*(t_k; x_{k,j}; L) \le \tfrac{L-\mu}{\mu\kappa}(f(t_k; x_{k,j}) - f^*(t_k)) \le 2\tfrac{L-\mu}{\mu\kappa} e^{-\sigma \cdot j}\Delta_k < \Delta_{k+1}.$$

That is a contradiction with the definition of Δ_{k+1}. □

COROLLARY 2.3.3

$$\sum_{k=0}^{N} j(k) \le (N+1)\left[1 + \sqrt{\tfrac{L}{\mu}} \cdot \ln \tfrac{2(L-\mu)}{\kappa\mu}\right] + \sqrt{\tfrac{L}{\mu}} \cdot \ln \tfrac{\Delta_0}{\Delta_{N+1}}.$$

□

It remains to estimate the number of internal iterations in the last step of the master process. Denote this number by j^*.

LEMMA 2.3.11

$$j^* \le 1 + \sqrt{\tfrac{L}{\mu}} \cdot \ln \tfrac{2(L-\mu)\Delta_{N+1}}{\kappa\mu\epsilon}.$$

Proof: The proof is very similar to that of Lemma 2.3.10. Suppose that

$$j^* - 1 > \sqrt{\tfrac{L}{\mu}} \cdot \ln \tfrac{2(L-\mu)\Delta_{N+1}}{\kappa\mu\epsilon}.$$

Note that for $j = j^* - 1$ we have

$$\begin{aligned} \epsilon &\le f^*(t_{N+1}; x_{N+1,j}; L) \le \tfrac{L-\mu}{\mu\kappa}(f(t_{N+1}; x_{N+1,j}) - f^*(t_{N+1})) \\ &\le 2\tfrac{L-\mu}{\mu\kappa} e^{-\sigma \cdot j}\Delta_{N+1} < \epsilon. \end{aligned}$$

That is a contradiction. □

COROLLARY 2.3.4

$$j^* + \sum_{k=0}^{N} j(k) \leq (N+2)\left[1 + \sqrt{\tfrac{L}{\mu}} \cdot \ln \tfrac{2(L-\mu)}{\kappa\mu}\right] + \sqrt{\tfrac{L}{\mu}} \cdot \ln \tfrac{\Delta_0}{\epsilon}.$$

Let us put all things together. Substituting estimate (2.3.24) for the number of full iterations N into the estimate of Corollary 2.3.4, we come to the following bound for the total number of internal iterations in the process (2.3.22):

$$\begin{aligned} &\left[\tfrac{1}{\ln[2(1-\kappa)]} \ln \tfrac{t_0 - t^*}{(1-\kappa)\epsilon} + 2\right] \cdot \left[1 + \sqrt{\tfrac{L}{\mu}} \cdot \ln \tfrac{2(L-\mu)}{\kappa\mu}\right] \\ &+ \sqrt{\tfrac{L}{\mu}} \cdot \ln \left(\tfrac{1}{\epsilon} \cdot \max_{1 \leq i \leq m} \{ f_0(x_0) - t_0; f_i(x_0) \} \right). \end{aligned} \tag{2.3.26}$$

Note that method (2.3.13), which implements the internal process, calls the oracle of problem (2.3.16) at each iteration only once. Therefore, we conclude that estimate (2.3.26) is an upper complexity bound for the problem (2.3.16) with ϵ-solution defined by (2.3.23). Let us check, how far this estimate is from the lower bounds.

The principal term in estimate (2.3.26) is of the order

$$\ln \tfrac{t_0 - t^*}{\epsilon} \cdot \sqrt{\tfrac{L}{\mu}} \cdot \ln \tfrac{L}{\mu}.$$

This value differs from the *lower bound* for an unconstrained minimization problem by a factor of $\ln \frac{L}{\mu}$. This means, that scheme (2.3.22) is at least *suboptimal* for constrained optimization problems. We cannot say more since a specific lower complexity bound for constrained minimization is not known.

To conclude this section, let us answer two technical questions. Firstly, in scheme (2.3.22) we assume that we know some estimate $t_0 < t^*$. This assumption is not binding since we can choose t_0 equal to the optimal value of the minimization problem

$$\min_{x \in Q} \left[f(x_0) + \langle f'(x_0), x - x_0 \rangle + \tfrac{\mu}{2} \parallel x - x_0 \parallel^2 \right].$$

Clearly, this value is less than or equal to t^*.

Secondly, we assume that we are able to compute $t^*(\bar{x}, t)$. Recall that $t^*(\bar{x}, t)$ is the root of function

$$f^*(t; \bar{x}; \mu) = \min_{x \in Q} f_\mu(t; \bar{x}; x),$$

where $f_\mu(t;\bar{x};x)$ is a max-type function composed by the components

$$f_0(\bar{x}) + \langle f_0'(\bar{x}), x - \bar{x}\rangle + \tfrac{\mu}{2} \parallel x - \bar{x} \parallel^2 - t,$$

$$f_i(\bar{x}) + \langle f_i'(\bar{x}), x - \bar{x}\rangle + \tfrac{\mu}{2} \parallel x - \bar{x} \parallel^2, \; i = 1 \dots m.$$

In view of Lemma 2.3.4, it is the optimal value of the following minimization problem:

$$\begin{array}{rl} & \min \; [f_0(\bar{x}) + \langle f_0'(\bar{x}), x - \bar{x}\rangle + \tfrac{\mu}{2} \parallel x - \bar{x} \parallel^2], \\ \text{s.t.} & f_i(\bar{x}) + \langle f_i'(\bar{x}), x - \bar{x}\rangle + \tfrac{\mu}{2} \parallel x - \bar{x} \parallel^2 \leq 0, \; i = 1 \dots m, \\ & x \in Q. \end{array}$$

This problem is not a quadratic optimization problem, since the constraints are not linear. However, it can be solved in finite time by a simplex-type process, since the objective function and the constraints have the same Hessian. This problem can be also solved by interior-point methods.

Chapter 3

NONSMOOTH CONVEX OPTIMIZATION

3.1 General convex functions

(Equivalent definitions; Closed functions; Continuity of convex functions; Separation theorems; Subgradients; Computation rules; Optimality conditions.)

3.1.1 Motivation and definitions

In this chapter we consider methods for solving general *convex* minimization problem

$$\begin{aligned} &\min\ f_0(x), \\ \text{s.t.}\ \ &f_i(x) \leq 0,\ i = 1 \dots m, \\ &x \in Q \subseteq R^n, \end{aligned} \tag{3.1.1}$$

where Q is a closed convex set and $f_i(x)$, $i = 0 \dots m$, are *general convex* functions. The term *general* means that these functions can be non-differentiable. Clearly, such a problem is more difficult than a smooth one.

Note that nonsmooth minimization problems arise frequently in different applications. Quite often some components of a model are composed by max-type functions:

$$f(x) = \max_{1 \leq j \leq p} \phi_j(x),$$

where $\phi_j(x)$ are convex and differentiable. In the previous section we have seen that such a function can be treated by a gradient mapping. However, if the number of smooth components p in this function is *very big*, the computation of the gradient mapping becomes too expensive.

Then, it is reasonable to treat this max-type function as a general convex function. Another source of nondifferentiable functions is the situation when some components of the problem (3.1.1) are given *implicitly*, as a solution of an auxiliary problem. Such functions are called the functions with *implicit* structure. Very often these functions appear to be nonsmooth.

Let us start our considerations with definition of general convex function. In the sequel the term "general" is often omitted.

Denote by

$$\operatorname{dom} f = \{x \in R^n : | f(x) | < \infty\}$$

the *domain* of function f. We always assume that $\operatorname{dom} f \neq \emptyset$.

DEFINITION 3.1.1 *Function $f(x)$ is called* convex *if its domain is convex and for all x, $y \in \operatorname{dom} f$ and $\alpha \in [0, 1]$ the following inequality holds:*

$$f(\alpha x + (1-\alpha)y) \leq \alpha f(x) + (1-\alpha)f(y).$$

We call f concave *if $-f$ is convex.*

At this point, we are not ready to speak about any method for solving (3.1.1). In the previous chapter, our optimization methods were based on the gradients of smooth functions. For nonsmooth functions such objects do not exist and we have to find something to replace them. However, in order to do that, we should study first the properties of general convex functions and justify a possibility to define a generalized gradient. That is a long way, but we have to pass through it.

A straightforward consequence of Definition 3.1.1 is as follows.

LEMMA 3.1.1 (Jensen inequality) *For any $x_1, \dots, x_m \in \operatorname{dom} f$ and coefficients $\alpha_1, \dots, \alpha_m$ such that*

$$\sum_{i=1}^{m} \alpha_i = 1, \quad \alpha_i \geq 0, \; i = 1 \dots m, \tag{3.1.2}$$

we have

$$f\left(\sum_{i=1}^{m} \alpha_i x_i\right) \leq \sum_{i=1}^{m} \alpha_i f(x_i).$$

Proof: Let us prove this statement by induction in m. Definition 3.1.1 justifies the inequality for $m = 2$. Assume it is true for some $m \geq 2$. For the set of $m+1$ points we have

$$\sum_{i=1}^{m+1} \alpha_i x_i = \alpha_1 x_1 + (1-\alpha_1) \sum_{i=1}^{m} \beta_i x_i,$$

where $\beta_i = \frac{\alpha_{i+1}}{1-\alpha_1}$. Clearly,

$$\sum_{i=1}^{m} \beta_i = 1, \quad \beta_i \geq 0, \; i = 1 \dots m.$$

Therefore, using Definition 3.1.1 and our inductive assumption, we have

$$\begin{aligned} f\left(\sum_{i=1}^{m+1} \alpha_i x_i\right) &= f\left(\alpha_1 x_1 + (1-\alpha_1) \sum_{i=1}^{m} \beta_i x_i\right) \\ &\leq \alpha_1 f(x_1) + (1-\alpha_1) f\left(\sum_{i=1}^{m} \beta_i x_i\right) \leq \sum_{i=1}^{m+1} \alpha_i f(x_i). \end{aligned}$$

□

The point $x = \sum_{i=1}^{m} \alpha_i x_i$ with coefficients α_i satisfying (3.1.2) is called a *convex combination* of points x_i.

Let us point out two important consequences of Jensen inequality.

COROLLARY 3.1.1 *Let x be a convex combination of points $x_1, \dots, x_m$. Then*

$$f(x) \leq \max_{1 \leq i \leq m} f(x_i).$$

Proof: Indeed, in view of Jensen inequality and since $\alpha_i \geq 0$, $\sum_{i=1}^{m} \alpha_i = 1$, we have

$$f(x) = f\left(\sum_{i=1}^{m} \alpha_i x_i\right) \leq \sum_{i=1}^{m} \alpha_i f(x_i) \leq \max_{1 \leq i \leq m} f(x_i).$$

□

COROLLARY 3.1.2 *Let*

$$\Delta = \text{Conv}\{x_1, \dots, x_m\} \equiv \{x = \sum_{i=1}^{m} \alpha_i x_i \mid \alpha_i \geq 0, \sum_{i=1}^{m} \alpha_i = 1\}.$$

Then $\max_{x \in \Delta} f(x) = \max_{1 \leq i \leq n} f(x_i)$. □

Let us give two equivalent definitions of convex functions.

THEOREM 3.1.1 *Function f is convex if and only if for all $x, y \in \text{dom} f$ and $\beta \geq 0$ such that $y + \beta(y - x) \in \text{dom} f$, we have*

$$f(y + \beta(y - x)) \geq f(y) + \beta(f(y) - f(x)). \qquad (3.1.3)$$

Proof: Let f be convex. Denote $\alpha = \frac{\beta}{1+\beta}$ and $u = y + \beta(y - x)$. Then

$$y = \tfrac{1}{1+\beta}(u + \beta x) = (1-\alpha)u + \alpha x.$$

Therefore

$$f(y) \le (1-\alpha)f(u) + \alpha f(x) = \tfrac{1}{1+\beta} f(u) + \tfrac{\beta}{1+\beta} f(x).$$

Let (3.1.3) hold. Let us fix $x, y \in \operatorname{dom} f$ and $\alpha \in (0,1]$. Denote $\beta = \frac{1-\alpha}{\alpha}$ and $u = \alpha x + (1-\alpha)y$. Then

$$x = \tfrac{1}{\alpha}(u - (1-\alpha)y) = u + \beta(u - y).$$

Therefore

$$f(x) \ge f(u) + \beta(f(u) - f(y)) = \tfrac{1}{\alpha} f(u) - \tfrac{1-\alpha}{\alpha} f(y).$$

□

THEOREM 3.1.2 *Function f is convex if and only if its* epigraph

$$\operatorname{epi}(f) = \{(x,t) \in \operatorname{dom} f \times R \mid \ t \ge f(x)\}$$

is a convex set.

Proof: Indeed, if $(x_1, t_1) \in \operatorname{epi}(f)$ and $(x_2, t_2) \in \operatorname{epi}(f)$, then for any $\alpha \in [0,1]$ we have

$$\alpha t_1 + (1-\alpha)t_2 \ge \alpha f(x_1) + (1-\alpha) f(x_2) \ge f(\alpha x_1 + (1-\alpha)x_2).$$

Thus, $(\alpha x_1 + (1-\alpha)x_2, \alpha t_1 + (1-\alpha)t_2) \in \operatorname{epi}(f)$.

Let $\operatorname{epi}(f)$ be convex. Note that for $x_1, x_2 \in \operatorname{dom} f$

$$(x_1, f(x_1)) \in \operatorname{epi}(f), \quad (x_1, f(x_2) \in \operatorname{epi}(f).$$

Therefore $(\alpha x_1 + (1-\alpha)x_2, \alpha f(x_1) + (1-\alpha)f(x_2)) \in \operatorname{epi}(f)$. That is

$$f(\alpha x_1 + (1-\alpha)x_2) \le \alpha f(x_1) + (1-\alpha) f(x_2).$$

□

We need also the following property of level sets of convex functions.

THEOREM 3.1.3 *If function f is convex, then all its level sets*

$$\mathcal{L}_f(\beta) = \{x \in \operatorname{dom} f \mid \ f(x) \le \beta\}$$

are either convex or empty.

Proof: Indeed, if $x_1 \in \mathcal{L}_f(\beta)$ and $x_2 \in \mathcal{L}_f(\beta)$, then for any $\alpha \in [0,1]$ we have

$$f(\alpha x_1 + (1-\alpha)x_2) \le \alpha f(x_1) + (1-\alpha) f(x_2) \le \alpha\beta + (1-\alpha)\beta = \beta.$$

□

We will see that the behavior of a general convex function on the boundary of its domain sometimes is out of any control. Therefore, let us introduce one convenient notion, which will be very useful in our analysis.

DEFINITION 3.1.2 *A convex function f is called* closed *if its epigraph is a closed set.*

As an immediate consequence of the definition we have the following result.

THEOREM 3.1.4 *If convex function f is closed, then all its level sets are either empty or closed.*

Proof: By its definition, $(\mathcal{L}_f(\beta), \beta) = \text{epi}\,(f) \bigcap \{(x,t) \mid t = \beta\}$. Therefore, the epigraph $\mathcal{L}_f(\beta)$ is closed and convex as an intersection of two closed convex sets. □

Note that, if f is convex and continuous and its domain $\text{dom}\, f$ is closed, then f is a closed function. However, in general, a closed convex function is not necessarily continuous.

Let us look at some examples of convex functions.

EXAMPLE 3.1.1 1. Linear function is closed and convex.

2. $f(x) = \mid x \mid$, $x \in R^1$, is closed and convex since its epigraph is

$$\{(x,t) \mid t \ge x,\ t \ge -x\},$$

the intersection of two closed convex sets (see Theorem 3.1.2).

3. All *differentiable* and convex on R^n functions belong to the class of general closed convex functions.

4. Function $f(x) = \frac{1}{x}$, $x > 0$, is convex and closed. However, its domain $\text{dom}\, f = \text{int}\, R^1_+$ is open.

5. Function $f(x) = \| x \|$, where $\| \cdot \|$ is any *norm*, is closed and convex:

$$\begin{aligned} f(\alpha x_1 + (1-\alpha)x_2) &= \| \alpha x_1 + (1-\alpha)x_2 \| \\ &\leq \| \alpha x_1 \| + \| (1-\alpha)x_2 \| \\ &= \alpha \| x_1 \| + (1-\alpha) \| x_2 \| \end{aligned}$$

for any $x_1, x_2 \in R^n$ and $\alpha \in [0,1]$. The most important norms in numerical analysis are so-called l_p-*norms*:

$$\| x \|_p = \left[\sum_{i=1}^{n} | x^{(i)} |^p \right]^{1/p}, \quad p \geq 1.$$

Among them there are three norms, which are commonly used:

- The *Euclidean norm*: $\| x \| = [\sum_{i=1}^{n} (x^{(i)})^2]^{1/2}$, $p = 2$.
- The l_1-*norm*: $\| x \|_1 = \sum_{i=1}^{n} | x^{(i)} |$, $p = 1$.
- The l_∞-*norm* (*Chebyshev* norm, *uniform* norm, *infinity* norm):

$$\| x \|_\infty = \max_{1 \leq i \leq n} | x^{(i)} |.$$

Any norm defines a system of *balls*,

$$B_{\|\cdot\|}(x_0, r) = \{ x \in R^n \mid \| x - x_0 \| \leq r \}, \quad r \geq 0,$$

where r is the *radius* of the ball and $x_0 \in R^n$ is its *center*. We call the ball $B_{\|\cdot\|}(0,1)$ the *unit* ball of the norm $\| \cdot \|$. Clearly, these balls are convex sets (see Theorem 3.1.3). For l_p-balls of the radius r we use the notation

$$B_p(x_0, r) = \{ x \in R^n \mid \| x - x_0 \|_p \leq r \}.$$

Note the following relation between Euclidean and l_1-balls:

$$B_1(x_0, r) \subset B_2(x_0, r) \subset B_1(x_0, r\sqrt{n}).$$

That is true because of the standard inequalities:

$$\sum_{i=1}^{n} (x^{(i)})^2 \leq \left(\sum_{i=1}^{n} | x^{(i)} | \right)^2,$$

$$\left(\frac{1}{n} \sum_{i=1}^{n} | x^{(i)} | \right)^2 \leq \frac{1}{n} \sum_{i=1}^{n} | x^{(i)} |^2.$$

6. Up to now, all our examples did not show up any pathological behavior. However, let us look at the following function of two variables:

$$f(x,y) = \begin{cases} 0, & \text{if } x^2+y^2<1, \\ \phi(x,y), & \text{if } x^2+y^2=1, \end{cases}$$

where $\phi(x,y)$ is an *arbitrary* nonnegative function defined on a unit circle. The domain of this function is the unit Euclidean disk, which is closed and convex. Moreover, it is easy to see that f is convex. However, it has no reasonable properties on the boundary of its domain. Definitely, we want to exclude such functions from our considerations. That was the reason for introducing the notion of closed function. It is clear that $f(x,y)$ is not closed unless $\phi(x,y) \equiv 0$. □

3.1.2 Operations with convex functions

In the previous section we have seen several examples of convex functions. Let us describe a set of invariant operations, which allow us to create more complicated objects.

THEOREM 3.1.5 *Let functions f_1 and f_2 be closed and convex and let $\beta \geq 0$. Then all functions below are closed and convex:*

1. $f(x) = \beta f_1(x)$, $\mathrm{dom}\, f = \mathrm{dom}\, f_1$.

2. $f(x) = f_1(x) + f_2(x)$, $\mathrm{dom}\, f = (\mathrm{dom}\, f_1) \bigcap (\mathrm{dom}\, f_2)$.

3. $f(x) = \max\{f_1(x), f_2(x)\}$, $\mathrm{dom}\, f = (\mathrm{dom}\, f_1) \bigcap (\mathrm{dom}\, f_2)$.

Proof:

1. The first item is evident:

$$f(\alpha x_1 + (1-\alpha)x_2) \leq \beta(\alpha f_1(x_1) + (1-\alpha) f_1(x_2)).$$

2. For all $x_1, x_2 \in (\mathrm{dom}\, f_1) \bigcap (\mathrm{dom}\, f_2)$ and $\alpha \in [0,1]$ we have

$$f_1(\alpha x_1 + (1-\alpha)x_2) + f_2(\alpha x_1 + (1-\alpha)x_2)$$

$$\leq \alpha f_1(x_1) + (1-\alpha) f_1(x_2) + \alpha f_2(x_1) + (1-\alpha) f_2(x_2)$$

$$= \alpha(f_1(x_1) + f_2(x_1)) + (1-\alpha)(f_1(x_2) + f_2(x_2)).$$

Thus, $f(x)$ is convex. Let us prove that it is closed. Consider a sequence $\{(x_k, t_k)\} \subset \mathrm{epi}\,(f)$:

$$t_k \geq f_1(x_k) + f_2(x_k), \quad \lim_{k\to\infty} x_k = \bar{x} \in \mathrm{dom}\, f, \quad \lim_{k\to\infty} t_k = \bar{t}.$$

Since f_1 and f_2 are closed, we have

$$\inf \lim_{k\to\infty} f_1(x_k) \geq f_1(\bar{x}), \quad \inf \lim_{k\to\infty} f_2(x_k) \geq f_2(\bar{x}).$$

Therefore

$$\bar{t} = \lim_{k\to\infty} t_k \geq \inf \lim_{k\to\infty} f_1(x_k) + \inf \lim_{k\to\infty} f_2(x_k) \geq f(\bar{x}).$$

Thus, $(\bar{x}, \bar{t}) \in \text{epi} f$.[1]

3. The epigraph of function $f(x)$ is as follows:

$$\begin{aligned} \text{epi} f &= \{(x,t) \mid t \geq f_1(x)\ t \geq f_2(x)\ x \in (\text{dom} f_1) \bigcap (\text{dom} f_2)\} \\ &\equiv \text{epi} f_1 \bigcap \text{epi} f_2. \end{aligned}$$

Thus, epi f is closed and convex as an intersection of two closed convex sets. It remains to use Theorem 3.1.2. □

The following theorem demonstrates that convexity is an *affine-invariant* property.

THEOREM 3.1.6 *Let function $\phi(y)$, $y \in R^m$, be convex and closed. Consider a linear operator*

$$\mathcal{A}(x) = Ax + b : \quad R^n \to R^m.$$

Then $f(x) = \phi(\mathcal{A}(x))$ is a closed convex function with domain

$$\text{dom} f = \{x \in R^n \mid \mathcal{A}(x) \in \text{dom} \phi\}.$$

Proof: For x_1 and x_2 from dom f denote $y_1 = \mathcal{A}(x_1)$, $y_2 = \mathcal{A}(y_2)$. Then for $\alpha \in [0,1]$ we have

$$\begin{aligned} f(\alpha x_1 + (1-\alpha)x_2) &= \phi(\mathcal{A}(\alpha x_1 + (1-\alpha)x_2)) \\ &= \phi(\alpha y_1 + (1-\alpha)y_2) \\ &\leq \alpha\phi(y_1) + (1-\alpha)\phi(y_2) \\ &= \alpha f(x_1) + (1-\alpha)f(x_2). \end{aligned}$$

[1] It is important to understand, that a similar property for convex sets is *not valid*. Consider the following two-dimensional example: $Q_1 = \{(x,y) : y \geq \frac{1}{x}, x > 0\}$, $Q_2 = \{(x,y) : y = 0, x \leq 0\}$. Both of these sets are convex and closed. However, their sum $Q_1 + Q_2 = \{(x,y) : y > 0\}$ is convex and *open*.

Thus, $f(x)$ is convex. The closedness of its epigraph follows from continuity of the linear operator $\mathcal{A}(x)$. □

The next theorem is one of the main suppliers of convex functions with implicit structure.

THEOREM 3.1.7 *Let* Δ *be some set and*

$$f(x) = \sup_y \{\phi(y, x) \mid y \in \Delta\}.$$

Suppose that for any fixed $y \in \Delta$ *the function* $\phi(y, x)$ *is closed and convex in* x. *Then* $f(x)$ *is a closed and convex function with domain*

$$\operatorname{dom} f = \{x \in \bigcap_{y \in \Delta} \operatorname{dom} \phi(y, \cdot) \mid \exists \gamma : \phi(y, x) \leq \gamma \; \forall y \in \Delta\}. \qquad (3.1.4)$$

Proof: Indeed, if x belongs to the right-hand side of equation (3.1.4), then $f(x) < \infty$ and we conclude that $x \in \operatorname{dom} f$. If x does not belong to this set, then there exists a sequence $\{y_k\}$ such that $\phi(y_k, x) \to \infty$. Therefore x does not belong to $\operatorname{dom} f$.

Finally, it is clear that $(x, t) \in \operatorname{epi} f$ if and only if for all $y \in \Delta$ we have

$$x \in \operatorname{dom} \phi(y, \cdot), \quad t \geq \phi(y, x).$$

This means that

$$\operatorname{epi} f = \bigcap_{y \in \Delta} \operatorname{epi} \phi(y, \cdot).$$

Therefore f is convex and closed since each $\operatorname{epi} \phi(y, \cdot)$ is convex and closed. □

Now we are ready to look at more sophisticated examples of convex functions.

EXAMPLE 3.1.2 1. Function $f(x) = \max_{1 \leq i \leq n} \{x^{(i)}\}$ is closed and convex.

2. Let $\lambda = (\lambda^{(1)}, \ldots, \lambda^{(m)})$ and Δ be a set in R^m_+. Consider the function

$$f(x) = \sup_{\lambda \in \Delta} \sum_{i=1}^{m} \lambda^{(i)} f_i(x),$$

where f_i are closed and convex. In view of Theorem 3.1.5, the epigraphs of functions

$$\phi_\lambda(x) = \sum_{i=1}^{m} \lambda^{(i)} f_i(x)$$

are convex and closed. Thus, $f(x)$ is closed and convex in view of Theorem 3.1.7. Note that we did not assume anything about the structure of the set Δ.

3. Let Q be a convex set. Consider the function

$$\psi_Q(x) = \sup\{\langle g, x\rangle \mid g \in Q\}.$$

Function $\psi_Q(x)$ is called *support* function of the set Q. Note that $\psi_Q(x)$ is closed and convex in view of Theorem 3.1.7. This function is homogeneous of degree one:

$$\psi_Q(tx) = t\psi_Q(x), \quad x \in \operatorname{dom} Q, \ t \geq 0.$$

If the set Q is bounded then $\operatorname{dom}\psi_Q = R^n$.

4. Let Q be a set in R^n. Consider the function $\psi(g,\gamma) = \sup\limits_{y\in Q} \phi(y,g,\gamma)$, where

$$\phi(y,g,\gamma) = \langle g, y\rangle - \tfrac{\gamma}{2} \parallel y \parallel^2 .$$

The function $\psi(g,\gamma)$ is closed and convex in (g,γ) in view of Theorem 3.1.7. Let us look at its properties.

If Q is bounded, then $\operatorname{dom}\psi = R^{n+1}$. Consider the case $Q = R^n$. Let us describe the domain of ψ. If $\gamma < 0$, then for any $g \neq 0$ we can take $y_\alpha = \alpha g$. Clearly, along this sequence $\phi(y_\alpha, g, \gamma) \to \infty$ as $\alpha \to \infty$. Thus, $\operatorname{dom}\psi$ contains only points with $\gamma \geq 0$.

If $\gamma = 0$, the only possible value for g is zero since otherwise the function $\phi(y, g, 0)$ is unbounded.

Finally, if $\gamma > 0$ then the point maximizing $\phi(y,g,\gamma)$ with respect to y is $y^*(g,\gamma) = \frac{1}{\gamma} g$ and we get the following expression for ψ:

$$\psi(g,\gamma) = \tfrac{\|g\|^2}{2\gamma}.$$

Thus,

$$\psi(g,\gamma) = \begin{cases} 0, & \text{if } g = 0, \gamma = 0, \\ \frac{\|g\|^2}{2\gamma}, & \text{if } \gamma > 0, \end{cases}$$

with the domain $\operatorname{dom}\psi = (R^n \times \{\gamma > 0\}) \bigcup (0,0)$. Note that this is a convex set, which is neither closed nor open. Nevertheless, ψ is a closed convex function. At the same time, this function is not continuous at the origin:

$$\psi(\sqrt{\gamma} g, \gamma) \equiv \tfrac{1}{2} \parallel g \parallel^2, \quad \gamma \neq 0.$$

□

3.1.3 Continuity and differentiability

In the previous sections we have seen that the behavior of convex functions at the boundary points of its domain can be rather disappointing (see Examples 3.1.1(6), 3.1.2(4)). Fortunately, this is the only bad news about convex functions. In this section we will see that the structure of convex functions in the *interior* of its domain is very simple.

LEMMA 3.1.2 *Let function f be convex and $x_0 \in \mathrm{int}\,(\mathrm{dom}\, f)$. Then f is locally upper bounded at x_0.*

Proof: Let us choose some $\epsilon > 0$ such that $x_0 \pm \epsilon e_i \in \mathrm{int}\,(\mathrm{dom}\, f)$, $i = 1 \dots n$, where e_i are the coordinate vectors of R^n. Denote

$$\Delta = \mathrm{Conv}\,\{x_0 \pm \epsilon e_i,\ i = 1 \dots n\}.$$

Let us show that $\Delta \supset B_2(x_0, \bar{\epsilon})$ with $\bar{\epsilon} = \frac{\epsilon}{\sqrt{n}}$. Indeed, consider

$$x = x_0 + \sum_{i=1}^{n} h_i e_i, \quad \sum_{i=1}^{n} (h_i)^2 \le \bar{\epsilon}.$$

We can assume that $h_i \ge 0$ (otherwise, in the above representation we can choose $-e_i$ instead of e_i). Then

$$\beta \equiv \sum_{i=1}^{n} h_i \le \sqrt{n} \sum_{i=1}^{n} (h_i)^2 \le \epsilon.$$

Therefore for $\bar{h}_i = \frac{1}{\beta} h_i$ we have

$$\begin{aligned} x &= x_0 + \beta \sum_{i=1}^{n} \bar{h}_i e_i = x_0 + \frac{\beta}{\epsilon} \sum_{i=1}^{n} \bar{h}_i \epsilon e_i \\ &= \left(1 - \frac{\beta}{\epsilon}\right) x_0 + \frac{\beta}{\epsilon} \sum_{i=1}^{n} \bar{h}_i (x_0 + \epsilon e_i) \in \Delta. \end{aligned}$$

Thus, using Corollary 3.1.2, we obtain

$$M \equiv \max_{x \in B_2(x_0, \bar{\epsilon})} f(x) \le \max_{x \in \Delta} f(x) \le \max_{1 \le i \le n} f(x_0 \pm \epsilon e_i).$$

□

Remarkably enough, the above result implies continuity of a convex function at any interior point of its domain.

THEOREM 3.1.8 *Let f be convex and $x_0 \in \mathrm{int}\,(\mathrm{dom}\, f)$. Then f is locally Lipschitz continuous at x_0.*

Proof: Let $B_2(x_0, \epsilon) \subseteq \mathrm{dom}\, f$ and $\sup\{f(x) \mid x \in B_2(x_0, \epsilon)\} \leq M$ (M is finite in view of Lemma 3.1.2). Consider $y \in B_2(x_0, \epsilon)$, $y \neq x_0$. Denote

$$\alpha = \tfrac{1}{\epsilon} \parallel y - x_0 \parallel, \quad z = x_0 + \tfrac{1}{\alpha}(y - x_0).$$

It is clear that $\parallel z - x_0 \parallel = \frac{1}{\alpha} \parallel y - x_0 \parallel = \epsilon$. Therefore $\alpha \leq 1$ and $y = \alpha z + (1 - \alpha)x_0$. Hence,

$$\begin{aligned} f(y) &\leq \alpha f(z) + (1-\alpha) f(x_0) \leq f(x_0) + \alpha(M - f(x_0)) \\ &= f(x_0) + \tfrac{M - f(x_0)}{\epsilon} \parallel y - x_0 \parallel . \end{aligned}$$

Further, denote $u = x_0 + \frac{1}{\alpha}(x_0 - y)$. Then $\parallel u - x_0 \parallel = \epsilon$ and $y = x_0 + \alpha(x_0 - u)$. Therefore, in view of Theorem 3.1.1 we have

$$\begin{aligned} f(y) &\geq f(x_0) + \alpha(f(x_0) - f(u)) \geq f(x_0) - \alpha(M - f(x_0)) \\ &= f(x_0) - \tfrac{M - f(x_0)}{\epsilon} \parallel y - x_0 \parallel . \end{aligned}$$

Thus, $\mid f(y) - f(x_0) \mid \leq \frac{M - f(x_0)}{\epsilon} \parallel y - x_0 \parallel$. □

Let us show that the convex functions possess a property, which is very close to differentiability.

DEFINITION 3.1.3 *Let $x \in \mathrm{dom}\, f$. We call f* differentiable in a direction *p at point x if the following limit exists:*

$$f'(x; p) = \lim_{\alpha \downarrow 0} \tfrac{1}{\alpha}[f(x + \alpha p) - f(x)]. \qquad (3.1.5)$$

The value $f'(x;p)$ is called the directional derivative *of f at x.*

THEOREM 3.1.9 *Convex function f is differentiable in any direction at any interior point of its domain.*

Proof: Let $x \in \mathrm{int}\,(\mathrm{dom}\, f)$. Consider the function

$$\phi(\alpha) = \tfrac{1}{\alpha}[f(x + \alpha p) - f(x)], \quad \alpha > 0.$$

Let $\gamma \in (0, 1]$ and $\alpha \in (0, \epsilon]$ be small enough to have $x + \epsilon p \in \mathrm{dom}\, f$. Then

$$f(x + \alpha\beta p) = f((1 - \beta)x + \beta(x + \alpha p)) \leq (1 - \beta) f(x) + \beta f(x + \alpha p).$$

Therefore

$$\phi(\alpha\beta) = \tfrac{1}{\alpha\beta}[f(x + \alpha\beta p) - f(x_0)] \leq \tfrac{1}{\alpha}[f(x + \alpha p) - f(x)] = \phi(\alpha).$$

Thus, $\phi(\alpha)$ decreases as $\alpha \downarrow 0$. Let us choose $\gamma > 0$ small enough to have $x - \gamma p \in \operatorname{dom} f$. Then, in view of (3.1.3) we have

$$\phi(\alpha) \geq \frac{1}{\gamma}[f(x) - f(x - \gamma p)].$$

Hence, the limit in (3.1.5) exists. □

Let us prove that the directional derivative provides us with a global lower support of the convex function.

LEMMA 3.1.3 *Let f be a convex function and $x \in \operatorname{int}(\operatorname{dom} f)$. Then $f'(x;p)$ is a convex function of p, which is homogeneous of degree 1. For any $y \in \operatorname{dom} f$ we have*

$$f(y) \geq f(x) + f'(x; y - x). \tag{3.1.6}$$

Proof: Let us prove that the directional derivative is homogeneous. Indeed, for $p \in R^n$ and $\tau > 0$ we have

$$\begin{aligned} f'(x;\tau p) &= \lim_{\alpha \downarrow 0} \frac{1}{\alpha}[f(x + \tau\alpha p) - f(x)] \\ &= \tau \lim_{\beta \downarrow 0} \frac{1}{\beta}[f(x + \beta p) - f(x)] = \tau f'(x_0; p). \end{aligned}$$

Further, for any p_1, $p_2 \in R^n$ and $\beta \in [0, 1]$ we obtain

$$\begin{aligned} f'(x; \beta p_1 + (1-\beta)p_2) &= \lim_{\alpha \downarrow 0} \frac{1}{\alpha}[f(x + \alpha(\beta p_1 + (1-\beta)p_2)) - f(x)] \\ &\leq \lim_{\alpha \downarrow 0} \frac{1}{\alpha}\{\beta[f(x + \alpha p_1) - f(x)] \\ &\quad +(1-\beta)[f(x + \alpha p_2) - f(x)]\} \\ &= \beta f'(x; p_1) + (1-\beta) f'(x; p_2). \end{aligned}$$

Thus, $f'(x;p)$ is convex in p. Finally, let $\alpha \in (0, 1]$, $y \in \operatorname{dom} f$ and $y_\alpha = x + \alpha(y - x)$. Then in view of Theorem 3.1.1, we have

$$f(y) = f(y_\alpha + \tfrac{1}{\alpha}(1-\alpha)(y_\alpha - x)) \geq f(y_\alpha) + \tfrac{1}{\alpha}(1-\alpha)[f(y_\alpha) - f(x)],$$

and we get (3.1.6) taking the limit in $\alpha \downarrow 0$. □

3.1.4 Separation theorems

Up to now we were describing properties of convex functions in terms of function values. We did not introduce any *directions* which could be useful for constructing the minimization schemes. In convex analysis such directions are defined by *separation theorems*, which form the subject of this section.

DEFINITION 3.1.4 *Let Q be a convex set. We say that hyperplane*

$$\mathcal{H}(g,\gamma) = \{x \in R^n \mid \langle g, x\rangle = \gamma\}, \quad g \neq 0,$$

is supporting *to Q if any $x \in Q$ satisfies inequality $\langle g, x\rangle \leq \gamma$.*

We say that the hyperplane $\mathcal{H}(g,\gamma)$ separates *a point x_0 from Q if*

$$\langle g, x\rangle \leq \gamma \leq \langle g, x_0\rangle \qquad (3.1.7)$$

for all $x \in Q$. If the second inequality in (3.1.7) is strict, we call the separation strict.

The separation theorems can be derived from the properties of projection.

DEFINITION 3.1.5 *Let Q be a closed set and $x_0 \in R^n$. Denote*

$$\pi_Q(x_0) = \arg\min\{\| x - x_0 \|: \ x \in Q\}.$$

We call $\pi_Q(x_0)$ projection *of point x_0 onto the set Q.*

THEOREM 3.1.10 *If Q is a convex set, then there exists a unique projection $\pi_Q(x_0)$.*

Proof: Indeed, $\pi_Q(x_0) = \arg\min\{\phi(x) \mid \ x \in Q\}$, where the function $\phi(x) = \frac{1}{2} \| x - x_0 \|^2$ belongs to $\mathcal{S}^{1,1}_{1,1}(R^n)$. Therefore $\pi_Q(x_0)$ is unique and well defined in view of Theorem 2.2.6. □

It is clear that $\pi_Q(x_0) = x_0$ if and only if $x_0 \in Q$.

LEMMA 3.1.4 *Let Q be a closed convex set and $x_0 \notin Q$. Then for any $x \in Q$ we have*

$$\langle \pi_Q(x_0) - x_0, x - \pi_Q(x_0)\rangle \geq 0. \qquad (3.1.8)$$

Proof: Note that $\pi_Q(x_0)$ is a solution to the minimization problem $\min\limits_{x\in Q} \phi(x)$ with $\phi(x) = \frac{1}{2} \| x - x_0 \|^2$. Therefore, in view of Theorem 2.2.5 we have

$$\langle \phi'(\pi_Q(x_0)), x - \pi_Q(x_0)\rangle \geq 0$$

for all $x \in Q$. It remains to note that $\phi'(x) = x - x_0$. □

Finally, we need a kind of *triangle inequality* for projection.

LEMMA 3.1.5 *For any $x \in Q$ we have*

$$\| x - \pi_Q(x_0) \|^2 + \| \pi_Q(x_0) - x_0 \|^2 \leq \| x - x_0 \|^2 .$$

Proof: Indeed, in view of (3.1.8), we have

$$\begin{aligned} \| x - \pi_Q(x_0) \|^2 - \| x - x_0 \|^2 &= \langle x_0 - \pi_Q(x_0), 2x - \pi_Q(x_0) - x_0 \rangle \\ &\leq - \| x_0 - \pi_Q(x_0) \|^2 . \end{aligned}$$

□

Now we can prove the separation theorems. We will need two of them. The first one describes our possibilities in strict separation.

THEOREM 3.1.11 *Let Q be a closed convex set and $x_0 \notin Q$. Then there exists a hyperplane $\mathcal{H}(g, \gamma)$, which strictly separates x_0 from Q. Namely, we can take*

$$g = x_0 - \pi_Q(x_0) \neq 0, \quad \gamma = \langle x_0 - \pi_Q(x_0), \pi_Q(x_0) \rangle.$$

Proof: Indeed, in view of (3.1.8), for any $x \in Q$ we have

$$\begin{aligned} \langle x_0 - \pi_Q(x_0), x \rangle &\leq \langle x_0 - \pi_Q(x_0), \pi_Q(x_0) \rangle \\ &= \langle x_0 - \pi_Q(x_0), x_0 \rangle - \| x_0 - \pi_Q(x_0) \|^2 . \end{aligned}$$

□

Let us give an example of an application of the above theorem.

COROLLARY 3.1.3 *Let Q_1 and Q_2 be two closed convex sets.*

1. If for any $g \in \operatorname{dom} \psi_{Q_2}$ we have $\psi_{Q_1}(g) \leq \psi_{Q_2}(g)$, then $Q_1 \subseteq Q_2$.

2. Let $\operatorname{dom} \psi_{Q_1} = \operatorname{dom} \psi_{Q_2}$ and for any $g \in \operatorname{dom} \psi_{Q_1}$ we have $\psi_{Q_1}(g) = \psi_{Q_2}(p)$. Then $Q_1 \equiv Q_2$.

Proof: 1. Assume that there exists $x_0 \in Q_1$, which does not belong to Q_2. Then, in view of Theorem 3.1.11, there exists a direction g such that

$$\langle g, x_0 \rangle > \gamma \geq \langle g, x \rangle$$

for all $x \in Q_2$. Hence, $g \in \operatorname{dom} \psi_{Q_2}$ and $\psi_{Q_1}(g) > \psi_{Q_2}(g)$. That is a contradiction.

2. In view of the first statement, $Q_1 \subseteq Q_2$ and $Q_2 \subseteq Q_1$. Therefore, $Q_1 \equiv Q_2$. □

The next separation theorem deals with boundary points of convex sets.

THEOREM 3.1.12 *Let Q be a closed convex set and x_0 belong to the boundary of set Q. Then there exists a hyperplane $\mathcal{H}(g,\gamma)$, supporting to Q and passing through x_0.*

(Such a vector g is called *supporting to Q at x_0*.)
Proof: Consider a sequence $\{y_k\}$ such that $y_k \notin Q$ and $y_k \to x_0$. Denote

$$g_k = \frac{y_k - \pi_Q(y_k)}{\|y_k - \pi_Q(y_k)\|}, \quad \gamma_k = \langle g_k, \pi_Q(y_k)\rangle.$$

In view of Theorem 3.1.11, for all $x \in Q$ we have

$$\langle g_k, x\rangle \leq \gamma_k \leq \langle g_k, y_k\rangle. \tag{3.1.9}$$

However, $\| g_k \| = 1$ and the sequence $\{\gamma_k\}$ is bounded:

$$\begin{aligned} \mid \gamma_k \mid &= \mid \langle g_k, \pi_Q(y_k) - x_0\rangle + \langle g_k, x_0\rangle \mid \\ \text{(Lemma 3.1.5)} \quad &\leq \| \pi_Q(y_k) - x_0 \| + \| x_0 \| \leq \| y_k - x_0 \| + \| x_0 \| . \end{aligned}$$

Therefore, without loss of generality we can assume that there exist $g^* = \lim_{k\to\infty} g_k$ and $\gamma^* = \lim_{k\to\infty} \gamma_k$. It remains to take the limit in (3.1.9). □

3.1.5 Subgradients

Now we are completely ready to introduce an extension of the notion of gradient.

DEFINITION 3.1.6 *Let f be a convex function. A vector g is called a* subgradient *of function f at point $x_0 \in \operatorname{dom} f$ if for any $x \in \operatorname{dom} f$ we have*

$$f(x) \geq f(x_0) + \langle g, x - x_0\rangle. \tag{3.1.10}$$

The set of all subgradients of f at x_0, $\partial f(x_0)$, is called the subdifferential *of function f at point x_0.*

The necessity of the notion of subdifferential is clear from the following example.

EXAMPLE 3.1.3 Consider function $f(x) = \mid x \mid$, $x \in R^1$. For all $y \in R^1$ and $g \in [-1, 1]$ we have

$$f(y) = \mid y \mid \geq g \cdot y = f(0) + g \cdot (y - 0).$$

Therefore, the subgradient of f at $x = 0$ is not unique. In our example it is the whole segment $[-1, 1]$. □

The whole set of inequalities (3.1.10), $x \in \mathrm{dom}\, f$, can be seen as a set of linear *constraints*, defining the set $\partial f(x_0)$. Therefore, by definition, the subdifferential is a *closed convex* set.

Note that the subdifferentiability of a function implies convexity.

LEMMA 3.1.6 *Let for any $x \in \mathrm{dom}\, f$ subdifferential $\partial f(x)$ be nonempty. Then f is a convex function.*

Proof: Indeed, let x, $y \in \mathrm{dom}\, f$, $\alpha \in [0, 1]$. Consider $y_\alpha = x + \alpha(y - x)$. Let $g \in \partial f(y_\alpha)$. Then

$$f(y) \geq f(y_\alpha) + \langle g, y - y_\alpha \rangle = f(y_\alpha) + (1 - \alpha)\langle g, y - x \rangle,$$

$$f(x) \geq f(y_\alpha) + \langle g, x - y_\alpha \rangle = f(y_\alpha) - \alpha \langle g, y - x \rangle.$$

Adding these inequalities multiplied by α and $(1 - \alpha)$ respectively, we get

$$\alpha f(y) + (1 - \alpha) f(x) \geq f(y_\alpha).$$

□

On the other hand, we can prove a converse statement.

THEOREM 3.1.13 *Let f be closed and convex and $x_0 \in \mathrm{int}\,(\mathrm{dom}\, f)$. Then $\partial f(x_0)$ is a nonempty bounded set.*

Proof: Note that the point $(f(x_0), x_0)$ belongs to the boundary of $\mathrm{epi}\,(f)$. Hence, in view of Theorem 3.1.12, there exists a hyperplane supporting to $\mathrm{epi}\,(f)$ at $(f(x_0), x_0)$:

$$-\alpha \tau + \langle d, x \rangle \leq -\alpha f(x_0) + \langle d, x_0 \rangle \tag{3.1.11}$$

for all $(\tau, x) \in \mathrm{epi}\,(f)$. Note that we can take

$$\| d \|^2 + \alpha^2 = 1. \tag{3.1.12}$$

Since for all $\tau \geq f(x_0)$ the point (τ, x_0) belongs to $\mathrm{epi}\,(f)$, we conclude that $\alpha \geq 0$.

Recall, that a convex function is locally upper bounded in the interior of its domain (Lemma 3.1.2). This means that there exist some $\epsilon > 0$ and $M > 0$ such that $B_2(x_0, \epsilon) \subseteq \mathrm{dom}\, f$ and

$$f(x) - f(x_0) \leq M \| x - x_0 \|$$

for all $x \in B_2(x_0, \epsilon)$. Therefore, in view of (3.1.11), for any x from this ball we have

$$\langle d, x - x_0 \rangle \leq \alpha(f(x) - f(x_0)) \leq \alpha M \parallel x - x_0 \parallel .$$

Choosing $x = x_0 + \epsilon d$ we get $\parallel d \parallel^2 \leq M\alpha \parallel d \parallel$. Thus, in view of the normalizing condition (3.1.12) we obtain

$$\alpha \geq \frac{1}{\sqrt{1+M^2}}.$$

Hence, choosing $g = d/\alpha$ we get

$$f(x) \geq f(x_0) + \langle g, x - x_0 \rangle$$

for all $x \in \operatorname{dom} f$.

Finally, if $g \in \partial f(x_0)$, $g \neq 0$, then choosing $x = x_0 + \epsilon g / \parallel g \parallel$ we obtain

$$\epsilon \parallel g \parallel = \langle g, x - x_0 \rangle \leq f(x) - f(x_0) \leq M \parallel x - x_0 \parallel = M\epsilon.$$

Thus, $\partial f(x_0)$ is bounded. □

Let us show that the conditions of the above theorem cannot be relaxed.

EXAMPLE 3.1.4 Consider the function $f(x) = -\sqrt{x}$ with the domain $\{x \in R^1 \mid x \geq 0\}$. This function is convex and closed, but the subdifferential does not exist at $x = 0$. □

Let us deterine an important relation between the subdifferential and the directional derivative of convex function.

THEOREM 3.1.14 *Let f be a closed convex function. For any $x_0 \in$ $\operatorname{int}(\operatorname{dom} f)$ and $p \in R^n$ we have*

$$f'(x_0; p) = \max\{\langle g, p \rangle \mid g \in \partial f(x_0)\}.$$

Proof: Note that

$$f'(x_0; p) = \lim_{\alpha \downarrow 0} \tfrac{1}{\alpha}[f(x_0 + \alpha p) - f(x_0)] \geq \langle g, p \rangle, \qquad (3.1.13)$$

where g is an arbitrary vector from $\partial f(x_0)$. Therefore, the subdifferential of function $f'(x_0; p)$ at $p = 0$ is not empty and $\partial f(x_0) \subseteq \partial_p f'(x_0; 0)$. On the other hand, since $f'(x_0;p)$ is convex in p, in view of Lemma 3.1.3, for any $y \in \mathrm{dom}\, f$ we have

$$f(y) \geq f(x_0) + f'(x_0; y - x_0) \geq f(x_0) + \langle g, y - x_0 \rangle,$$

where $g \in \partial_p f'(x_0; 0)$. Thus, $\partial_p f'(x_0;0) \subseteq \partial f(x_0)$ and we conclude that $\partial f(x_0) \equiv \partial_p f'(x_0; 0)$.

Consider $g_p \in \partial_p f'(x_0;p)$. Then, in view of inequality (3.1.6), for all $v \in R^n$ and $\tau > 0$ we have

$$\tau f'(x_0; v) = f'(x_0; \tau v) \geq f'(x_0;p) + \langle g_p, \tau v - p \rangle.$$

Considering $\tau \to \infty$ we conclude that

$$f'(x_0; v) \geq \langle g_p, v \rangle, \qquad (3.1.14)$$

and, considering $\tau \to 0$, we obtain

$$f'(x_0;p) - \langle g_p, p \rangle \leq 0. \qquad (3.1.15)$$

However, inequality (3.1.14) implies that $g_p \in \partial_p f'(x_0; 0)$. Therefore, comparing (3.1.13) and (3.1.15) we conclude that $\langle g_p, p \rangle = f'(x_0;p)$. □

To conclude this section, let us point out several properties of subgradients, which are of main importance for optimization. Let us start from the optimality condition.

THEOREM 3.1.15 *We have* $f(x^*) = \min\limits_{x \in \mathrm{dom}\, f} f(x)$ *if and only if*

$$0 \in \partial f(x^*).$$

Proof: Indeed, if $0 \in \partial f(x^*)$, then $f(x) \geq f(x^*) + \langle 0, x - x^* \rangle = f(x^*)$ for all $x \in \mathrm{dom}\, f$. On the other hand, if $f(x) \geq f(x^*)$ for all $x \in \mathrm{dom}\, f$, then $0 \in \partial f(x^*)$ in view of Definition 3.1.6. □

The next result forms a basis for *cutting plane* optimization schemes.

THEOREM 3.1.16 *For any* $x_0 \in \mathrm{dom}\, f$ *all vectors* $g \in \partial f(x_0)$ *are supporting to the level set* $\mathcal{L}_f(f(x_0))$*:*

$$\langle g, x_0 - x \geq 0 \quad \forall x \in \mathcal{L}_f(f(x_0)) \equiv \{x \in \mathrm{dom}\, f : \; f(x) \leq f(x_0)\}.$$

Proof: Indeed, if $f(x) \leq f(x_0)$ and $g \in \partial f(x_0)$, then

$$f(x_0) + \langle g, x - x_0 \rangle \leq f(x) \leq f(x_0).$$

□

COROLLARY 3.1.4 *Let $Q \subseteq \mathrm{dom}\, f$ be a closed convex set, $x_0 \in Q$ and*

$$x^* = \arg\min\{f(x) \mid x \in Q\}.$$

Then for any $g \in \partial f(x_0)$ we have $\langle g, x_0 - x^* \rangle \geq 0$. □

3.1.6 Computing subgradients

In the previous section we introduced the subgradients, objects which we are going to use in minimization schemes. However, in order to apply such schemes in practice, we need to be sure that these objects are computable. In this section we present some rules for computing the things.

LEMMA 3.1.7 *Let f be closed and convex. Assume that it is differentiable on its domain. Then $\partial f(x) = \{f'(x)\}$ for any $x \in \mathrm{int}\,(\mathrm{dom}\, f)$.*

Proof: Let us fix some $x \in \mathrm{int}\,(\mathrm{dom}\, f)$. Then, in view of Theorem 3.1.14, for any direction $p \in R^n$ and any $g \in \partial f(x)$ we have

$$\langle f'(x), p \rangle = f'(x; p) \geq \langle g, p \rangle.$$

Changing the sign of p, we conclude that $\langle f'(x), p \rangle = \langle g, p \rangle$ for all g from $\partial f(x)$. Finally, considering $p = e_k$, $k = 1 \ldots n$, we get $g = f'(x)$. □

Let us provide all operations with convex functions, described in Section 3.1.2, with corresponding rules for updating subgradients.

LEMMA 3.1.8 *Let function $f(y)$ be closed and convex with* $\mathrm{dom}\, f \subseteq R^m$. *Consider a linear operator*

$$\mathcal{A}(x) = Ax + b : \quad R^n \to R^m.$$

Then $\phi(x) = f(\mathcal{A}(x))$ is a closed convex function with domain $\mathrm{dom}\, \phi = \{x \mid \mathcal{A}(x) \in \mathrm{dom}\, f\}$. *For any $x \in \mathrm{int}\,(\mathrm{dom}\, \phi)$ we have*

$$\partial \phi(x) = A^T \partial f(\mathcal{A}(x)).$$

Proof: We have already proved the first part of this lemma in Theorem 3.1.6. Let us prove the relation for the subdifferential.

Indeed, let $y_0 = \mathcal{A}(x_0)$. Then for all $p \in R^n$ we have

$$\begin{aligned} \phi'(x_0, p) = f'(y_0; Ap) &= \max\{\langle g, Ap \rangle \mid g \in \partial f(y_0)\} \\ &= \max\{\langle \bar{g}, p \rangle \mid \bar{g} \in A^T \partial f(y_0)\}. \end{aligned}$$

Using Theorem 3.1.14 and Corollary 3.1.3, we get

$$\partial\phi(x_0) = A^T \partial f(\mathcal{A}(x_0)).$$

□

LEMMA 3.1.9 *Let $f_1(x)$ and $f_2(x)$ be closed convex functions and α_1, $\alpha_2 \geq 0$. Then function $f(x) = \alpha_1 f_1(x) + \alpha_2 f_2(x)$ is closed and convex and*

$$\partial f(x) = \alpha_1 \partial f_1(x) + \alpha_2 \partial f_2(x) \tag{3.1.16}$$

for any x from $\text{int}\,(\text{dom}\, f) = \text{int}\,(\text{dom}\, f_1) \bigcap \text{int}\,(\text{dom}\, f_2)$.

Proof: In view of Theorem 3.1.5, we need to prove only the relation for the subdifferentials. Consider $x_0 \in \text{int}\,(\text{dom}\, f_1) \bigcap \text{int}\,(\text{dom}\, f_2)$. Then, for any $p \in R^n$ we have

$$\begin{aligned} f'(x_0;p) &= \alpha_1 f_1'(x_0;p) + \alpha_2 f_2'(x_0;p) \\ &= \max\{\langle g_1, \alpha_1 p\rangle \mid g_1 \in \partial f_1(x_0)\} \\ &\quad + \max\{\langle g_2, \alpha_2 p\rangle \mid g_2 \in \partial f_2(x_0)\} \\ &= \max\{\langle \alpha_1 g_1 + \alpha_2 g_2, p\rangle \mid g_1 \in \partial f_1(x_0),\ g_2 \in \partial f_2(x_0)\} \\ &= \max\{\langle g, p\rangle \mid g \in \alpha_1 \partial f_1(x_0) + \alpha_2 \partial f_2(x_0)\}. \end{aligned}$$

Note that both $\partial f_1(x_0)$ and $\partial f_1(x_0)$ are bounded. Hence, using Theorem 3.1.14 and Corollary 3.1.3, we get (3.1.16). □

LEMMA 3.1.10 *Let functions $f_i(x)$, $i = 1 \ldots m$, be closed and convex. Then function $f(x) = \max\limits_{1 \leq i \leq m} f_i(x)$ is also closed and convex. For any $x \in \text{int}\,(\text{dom}\, f) = \bigcap\limits_{i=1}^{m} \text{int}\,(\text{dom}\, f_i)$ we have*

$$\partial f(x) = \text{Conv}\,\{\partial f_i(x) \mid i \in I(x)\}, \tag{3.1.17}$$

where $I(x) = \{i : f_i(x) = f(x)\}$.

Proof: Again, in view of Theorem 3.1.5, we need to justify only the rules for subdifferentials. Consider $x \in \bigcap\limits_{i=1}^{m} \text{int}\,(\text{dom}\, f_i)$. Assume that $I(x) = \{1, \ldots, k\}$. Then for any $p \in R^n$ we have

$$f'(x;p) = \max_{1 \leq i \leq k} f_i'(x;p) = \max_{1 \leq i \leq k} \max\{\langle g_i, p\rangle \mid g_i \in \partial f_i(x)\}.$$

Note that for any set of values $a_1, \dots, a_k$ we have

$$\max_{1 \le i \le k} a_i = \max \left\{ \sum_{i=1}^{k} \lambda_i a_i \mid \{\lambda_i\} \in \Delta_k \right\},$$

where $\Delta_k = \{\lambda_i \ge 0, \ \sum\limits_{i=1}^{k} \lambda_i = 1\}$, the k-dimensional *standard simplex*. Therefore,

$$\begin{aligned} f'(x;p) &= \max_{\{\lambda_i\} \in \Delta_k} \{ \sum_{i=1}^{k} \lambda_i \max\{\langle g_i, p\rangle \mid g_i \in \partial f_i(x)\} \\ &= \max\{ \langle \sum_{i=1}^{k} \lambda_i g_i, p\rangle \mid g_i \in \partial f_i(x), \ \{\lambda_i\} \in \Delta_k\} \\ &= \max\{\langle g, p\rangle \mid g = \sum_{i=1}^{k} \lambda_i g_i, g_i \in \partial f_i(x), \{\lambda_i\} \in \Delta_k\} \\ &= \max\{\langle g, p\rangle \mid g \in \operatorname{Conv}\{\partial f_i(x), i \in I(x)\} \ \}. \end{aligned}$$

□

The last rule can be useful for computing some elements from the subdifferential.

LEMMA 3.1.11 *Let Δ be a set and $f(x) = \sup\{\phi(y,x) \mid y \in \Delta\}$. Suppose that for any fixed $y \in \Delta$ the function $\phi(y,x)$ is closed and convex in x. Then $f(x)$ is closed convex.*

Moreover, for any x from

$$\operatorname{dom} f = \{x \in R^n \mid \exists \gamma : \phi(y,x) \le \gamma \ \forall y \in \Delta\}$$

we have

$$\partial f(x) \supseteq \operatorname{Conv}\{\partial \phi_x(y,x) \mid y \in I(x)\},$$

where $I(x) = \{y \mid \phi(y,x) = f(x)\}$.

Proof: In view of Theorem 3.1.7, we have to prove only the inclusion. Indeed, for any $x \in \operatorname{dom} f$, $y \in I(x)$ and $g \in \partial \phi_x(y,x)$ we have

$$f(x) \ge \phi(y,x) \ge \phi(y,x_0) + \langle g, x - x_0\rangle = f(x_0) + \langle g, x - x_0\rangle.$$

□

Now we can look at some examples of subdifferentials.

EXAMPLE 3.1.5 1. Let $f(x) =\mid x \mid$, $x \in R^1$. Then $\partial f(0) = [-1,1]$ since

$$f(x) = \max_{-1 \le g \le 1} g \cdot x.$$

2. Consider function $f(x) = \sum\limits_{i=1}^{m} \mid \langle a_i, x\rangle - b_i \mid$. Denote

$$\begin{aligned} I_-(x) &= \{i : \langle a_i, x\rangle - b_i < 0\}, \\ I_+(x) &= \{i : \langle a_i, x\rangle - b_i > 0\}, \\ I_0(x) &= \{i : \langle a_i, x\rangle - b_i = 0\}. \end{aligned}$$

   Then $\partial f(x) = \sum\limits_{i\in I_+(x)} a_i - \sum\limits_{i\in I_-(x)} a_i + \sum\limits_{i\in I_0(x)} [-a_i, a_i]$.

3. Consider function $f(x) = \max\limits_{1\le i\le n} x^{(i)}$. Denote $I(x) = \{i : x^{(i)} = f(x)\}$. Then $\partial f(x) = \text{Conv}\,\{e_i \mid i \in I(x)\}$. For $x = 0$ we have

$$\partial f(0) = \text{Conv}\,\{e_i \mid 1 \le i \le n\} \equiv \Delta_n.$$

4. For Euclidean norm $f(x) = \parallel x \parallel$ we have

$$\begin{aligned} \partial f(0) &= B_2(0,1) = \{x \in R^n \mid \parallel x \parallel \le 1\}, \\ \partial f(x) &= \{x / \parallel x \parallel\}, \; x \ne 0. \end{aligned}$$

5. For l_1-norm $f(x) = \parallel x \parallel_1 = \sum\limits_{i=1}^{n} \mid x^{(i)} \mid$ we have

$$\begin{aligned} \partial f(0) &= B_\infty(0,1) = \{x \in R^n \mid \max_{1\le i\le n} \mid x^{(i)} \mid \le 1\}, \\ \partial f(x) &= \sum_{i\in I_+(x)} e_i - \sum_{i\in I_-(x)} e_i + \sum_{i\in I_0(x)} [-e_i, e_i], \; x \ne 0, \end{aligned}$$

   where $I_+(x) = \{i \mid x^{(i)} > 0\}$, $I_-(x) = \{i \mid x^{(i)} < 0\}$ and $I_0(x) = \{i \mid x^{(i)} = 0\}$.

We leave justification of these examples as an exercise for the reader. □

We conclude this section with an example of application of the above technique for deriving an optimality condition for a smooth minimization problem with functional constraints.

THEOREM 3.1.17 *(Kuhn-Tucker). Let f_i be differentiable convex functions, $i = 0 \ldots m$. Suppose that there exists a point $\bar{x}$ such that $f_i(\bar{x}) < 0$ for all $i = 1 \ldots m$. (Slater condition.)*

A point x^* is a solution to the problem

$$\min\{f_0(x) \mid f_i(x) \leq 0, i = 1 \dots m\} \tag{3.1.18}$$

if and only if it is feasible and there exist nonnegative numbers λ_i, $i = 1 \dots m$, such that

$$f_0'(x^*) + \sum_{i \in I^*} \lambda_i f_i'(x^*) = 0,$$

where $I^* = \{i \in [1, m] : f_i(x^*) = 0\}$.

Proof: In view of Lemma 2.3.4, x^* is a solution to (3.1.18) if and only if it is a global minimizer of the function

$$\phi(x) = \max\{f_0(x) - f^*; f_i(x), i = 1 \dots m\}.$$

In view of Theorem 3.1.15, this is the case if and only if $0 \in \partial\phi(x^*)$. Further, in view of Lemma 3.1.10, this is true if and only if there exist nonnegative $\bar{\lambda}_i$, such that

$$\bar{\lambda}_0 f_0'(x^*) + \sum_{i \in I^*} \bar{\lambda}_i f_i'(x^*) = 0, \quad \bar{\lambda}_0 + \sum_{i \in I^*} \bar{\lambda}_i = 1.$$

Thus, we need to prove only that $\bar{\lambda}_0 > 0$. Indeed, if $\bar{\lambda}_0 = 0$, then

$$\sum_{i \in I^*} \bar{\lambda}_i f_i(\bar{x}) \geq \sum_{i \in I^*} \bar{\lambda}_i [f_i(x^*) + \langle f_i'(x^*), \bar{x} - x^* \rangle] = 0.$$

This contradicts the Slater condition. Therefore $\bar{\lambda}_0 > 0$ and we can take $\lambda_i = \bar{\lambda}_i / \bar{\lambda}_0$, $i \in I^*$. □

Theorem 3.1.17 is very useful for solving simple optimization problems.

LEMMA 3.1.12 *Let $A \succ 0$. Then*

$$\max_x \{\langle c, x \rangle : \langle Ax, x \rangle \leq 1\} = \langle A^{-1}c, c \rangle^{1/2}.$$

Proof: Note that all conditions of Theorem 3.1.17 are satisfied and the solution x^* of the above problem is attained at the boundary of the feasible set. Therefore, in accordance with Theorem 3.1.17 we have to solve the following equations:

$$c = \lambda A x^*, \quad \langle Ax^*, x^* \rangle = 1.$$

Thus, $\lambda = \langle A^{-1}c, c \rangle^{1/2}$ and $x^* = \frac{1}{\lambda} A^{-1} c$. □

3.2 Nonsmooth minimization methods

(General lower complexity bounds; Main lemma; Localization sets; Subgradient method; Constrained minimization scheme; Optimization in finite dimension and lower complexity bounds; Cutting plane scheme; Center of gravity method; Ellipsoid method; Other methods.)

3.2.1 General lower complexity bounds

In the previous section we have introduced a class of general convex functions. These functions can be nonsmooth and therefore the corresponding minimization problem can be quite difficult. As for smooth problems, let us try to derive a lower complexity bounds, which will help us to evaluate the performance of numerical methods.

In this section we derive such bounds for the following unconstrained minimization problem

$$\min_{x \in R^n} f(x), \tag{3.2.1}$$

where f is a convex function. Thus, our problem class is as follows:

Model:	1. Unconstrained minimization. 2. f is convex on R^n and Lipschitz continuous on a bounded set.
Oracle:	First-order black box: at each point $\hat{x}$ we can compute $f(\hat{x}), \quad g(\hat{x}) \in \partial f(\hat{x})$, $g(\hat{x})$ is an *arbitrary* subgradient.
Approximate solution:	Find $\bar{x} \in R^n : f(\bar{x}) - f^* \leq \epsilon$.
Methods:	Generate a sequence $\{x_k\}$: $x_k \in x_0 + \mathrm{Lin}\,\{g(x_0), \dots, g(x_{k-1})\}$.

(3.2.2)

As in Section 2.1.2, for deriving a lower complexity bound for our problem class, we will study the behavior of numerical methods on some function, which appears to be very difficult for all of them.

Let us fix some constants $\mu > 0$ and $\gamma > 0$. Consider the family of functions

$$f_k(x) = \gamma \max_{1\leq i\leq k} x^{(i)} + \tfrac{\mu}{2} \| x \|^2, \quad k = 1 \ldots n.$$

Using the rules of subdifferential calculus, described in Section 3.1.6, we can write down an expression for the subdifferential of f_k at x. That is

$$\begin{aligned} \partial f_k(x) &= \mu x + \gamma \mathrm{Conv}\,\{e_i \mid i \in I(x)\}, \\ I(x) &= \{j \mid 1 \leq j \leq k,\ x^{(j)} = \max_{1\leq i\leq k} x^{(i)}\}. \end{aligned}$$

Therefore for any x, $y \in B_2(0,\rho)$, $\rho > 0$, and $g_k(y) \in \partial f_k(y)$ we have

$$\begin{aligned} f_k(y) - f_k(x) &\leq \langle g_k(y), y - x\rangle \\ &\leq \| g_k(y) \| \cdot \| y - x \| \leq (\mu\rho + \gamma) \| y - x \| . \end{aligned}$$

Thus, f_k is Lipschitz continuous on $B_2(0,\rho)$ with Lipschitz constant $M = \mu\rho + \gamma$.

Further, consider the point x_k^* with the coordinates

$$(x_k^*)^{(i)} = \begin{cases} -\frac{\gamma}{\mu k}, & 1 \leq i \leq k, \\ 0, & k+1 \leq i \leq n. \end{cases}$$

It is easy to check that $0 \in \partial f_k(x_k^*)$ and therefore x_k^* is the minimum of function $f_k(x)$ (see Theorem 3.1.15). Note that

$$R_k \equiv \| x_k^* \| = \frac{\gamma}{\mu\sqrt{k}}, \quad f_k^* = -\frac{\gamma^2}{\mu k} + \frac{\mu}{2} R_k^2 = -\frac{\gamma^2}{2\mu k}.$$

Let us describe now a resisting oracle for function $f_k(x)$. Since the analytical form of this function is fixed, the resistance of this oracle consists in providing us with the worst possible subgradient at each test point. The algorithmic scheme of this oracle is as follows.

Input:	$x \in R^n$.
Main Loop:	$f := -\infty; \quad i^* := 0;$ **for** $j := 1$ **to** m **do** **if** $x^{(j)} > f$ **then** $\{ f := x^{(j)}; \; i^* := j \}$; $f := \gamma f + \frac{\mu}{2} \parallel x \parallel^2; \quad g := e_{i^*} + \mu x;$
Output:	$f_k(x) := f, \quad g_k(x) := g \in R^n$.

At the first glance, there is nothing special in this scheme. Its main loop is just a standard process for finding a maximal coordinate of a vector from R^n. However, the main feature of this loop is that we always form the subgradient as a coordinate vector. Moreover, this coordinate corresponds to i^*, which is the first maximal component of vector x. Let us check what happens with a minimizing sequence, which uses such an oracle.

Let us choose starting point $x_0 = 0$. Denote

$$R^{p,n} = \{x \in R^n \mid x^{(i)} = 0, \; p+1 \leq i \leq n\}.$$

Since $x_0 = 0$, the answer of the oracle is $f_k(x_0) = 0$ and $g_k(x_0) = e_1$. Therefore the next point of the sequence, x_1, necessarily belongs to $R^{1,n}$. Assume now that the current test point of the sequence, x_i, belongs to $R^{p,n}$, $1 \leq p \leq k$. Then the oracle will return a subgradient

$$g = \mu x_i + \gamma e_{i^*},$$

where $i^* \leq p + 1$. Therefore, the next test point x_{i+1} belongs to $R^{p+1,n}$.

This simple reasoning proves that for all i, $1 \leq i \leq k$, we have $x_i \in R^{i,n}$. Consequently, for i: $1 \leq i \leq k-1$, we cannot improve the starting value of the objective function:

$$f_k(x_i) \geq \gamma \max_{1 \leq j \leq k} x_i^{(j)} = 0.$$

Let us convert this observation in a lower complexity bound. Let us fix some parameters of our problem class $\mathcal{P}(x_0, R, M)$, that is $R > 0$ and $M > 0$. In addition to (3.2.2) we assume that

- the solution of problem (3.2.1), x^*, exists and $x^* \in B_2(x_0, R)$.
- f is Lipschitz continuous on $B_2(x_0, R)$ with constant $M > 0$.

THEOREM 3.2.1 *For any class $\mathcal{P}(x_0, R, M)$ and any k, $0 \le k \le n-1$, there exists a function $f \in \mathcal{P}(x_0, R, M)$ such that*

$$f(x_k) - f^* \ge \frac{MR}{2(1+\sqrt{k+1})}$$

for any optimization scheme, which generates a sequence $\{x_k\}$ satisfying the condition

$$x_k \in x_0 + \mathrm{Lin}\{g(x_0), \ldots, g(x_{k-1})\}.$$

Proof: Without loss of generality we can assume that $x_0 = 0$. Let us choose $f(x) = f_{k+1}(x)$ with

$$\gamma = \frac{\sqrt{k+1}M}{1+\sqrt{k+1}}, \quad \mu = \frac{M}{(1+\sqrt{k+1})R}.$$

Then

$$f^* = f^*_{k+1} = -\frac{\gamma^2}{2\mu(k+1)} = -\frac{MR}{2(1+\sqrt{k+1})},$$

$$\| x_0 - x^* \| = R_{k+1} = \frac{\gamma}{\mu\sqrt{k+1}} = R,$$

and $f(x)$ is Lipschitz continuous on $B_2(x_0, R)$ with constant $\mu R + \gamma = M$. Note that $x_k \in R^{k,n}$. Hence, $f(x_k) - f^* \ge -f^*$. □

The lower complexity bound presented in Theorem 3.2.1 is uniform in the dimension of the space of variables. As for the lower bound of Theorem 2.1.7, it can be applied to problems with very large dimension, or to efficiency analysis of starting iterations of a minimization scheme ($k \le n-1$).

We will see that our lower estimate is exact: There exist minimization methods, which have the rate of convergence proportional to this lower bound. Comparing this bound with the lower bound for smooth minimization problems, we can see that now the possible convergence rate is much slower. However, we should remember that we are working now with the most general class of convex problems.

3.2.2 Main lemma

At this moment we are interested in the following problem:

$$\min\{f(x) \mid x \in Q\}, \tag{3.2.3}$$

where Q is a closed convex set, and f is a function, which is convex on R^n. We are going to study some methods for solving (3.2.3), which employ subgradients $g(x)$ of the objective function. As compared with the smooth problem, our goal now is much more complicated. Indeed, even in the simplest situation, when $Q \equiv R^n$, the subgradient seems to be a poor replacement for the gradient of smooth function. For example, we cannot be sure that the value of the objective function is decreasing in the direction $-g(x)$. We cannot expect that $g(x) \to 0$ as x approaches a solution of our problem, etc.

Fortunately, there is one property of subgradients that makes our goals reachable. We have proved this property in Corollary 3.1.4:

At any $x \in Q$ the following inequality holds:

$$\langle g(x), x - x^* \rangle \geq 0. \tag{3.2.4}$$

This simple inequality leads to two consequences, which form a basis for any nonsmooth minimization method. Namely:

- The distance between x and x^* is decreasing in the direction $-g(x)$.
- Inequality (3.2.4) cuts R^n on two half-spaces. Only one of them contains x^*.

Nonsmooth minimization methods cannot employ the idea of relaxation or approximation. There is another concept, underlying all these schemes. That is the concept of *localization*. However, to go forward with this concept, we have to develop some special technique, which allows us to estimate the quality of an approximate solution to problem (3.2.3). That is the main goal of this section.

Let us fix some $\bar{x} \in R^n$. For $x \in R^n$ with $g(x) \neq 0$ define

$$v_f(\bar{x}, x) = \tfrac{1}{\| g(x) \|} \langle g(x), x - \bar{x} \rangle.$$

If $g(x) = 0$, then define $v_f(\bar{x}; x) = 0$. Clearly, $v_f(\bar{x}, x) \leq \| x - \bar{x} \|$.

The values $v_f(\bar{x}, x)$ have a natural geometric interpretation. Consider a point x such that $g(x) \neq 0$ and $\langle g(x), x - \bar{x} \rangle \geq 0$. Let us look at the point $y = \bar{x} + v_f(x) g(x) / \| g(x) \|$. Then

$$\langle g(x), x - y \rangle = \langle g(x), x - \bar{x} \rangle - v_f(\bar{x}, x) \| g(x) \| = 0$$

and $\| y - \bar{x} \| = v_f(\bar{x}, x)$. Thus, $v_f(\bar{x}, x)$ is a *distance* from the point $\bar{x}$ to hyperplane $\{y : \langle g(x), x - y \rangle = 0\}$.

Let us introduce a function that measures the variation of function f with respect to the point $\bar{x}$. For $t \geq 0$ define

$$\omega_f(\bar{x}; t) = \max\{ f(x) - f(\bar{x}) \mid \| x - \bar{x} \| \leq t \}.$$

If $t < 0$, we set $\omega_f(\bar{x}; t) = 0$.

Clearly, the function ω_f possesses the following properties:

- $\omega_f(\bar{x};0) = 0$ for all $t \le 0$.
- $\omega_f(\bar{x};t)$ is a nondecreasing function of t, $t \in R^1$.
- $f(x) - f(\bar{x}) \le \omega_f(\bar{x}; \| x - x^* \|)$.

It is important that in the convex situation the last inequality can be strengthened.

LEMMA 3.2.1 *For any $x \in R^n$ we have*

$$f(x) - f(\bar{x}) \le \omega_f(\bar{x}; v_f(\bar{x};x)). \qquad (3.2.5)$$

If $f(x)$ is Lipschitz continuous on $B_2(\bar{x},R)$ with some constant M, then

$$f(x) - f(\bar{x}) \le M(v_f(\bar{x};x))_+ \qquad (3.2.6)$$

for all $x \in R^n$ with $v_f(\bar{x};x) \le R$.

Proof: If $\langle g(x), x - \bar{x}\rangle \le 0$, then $f(\bar{x}) \ge f(x) + \langle g(x), \bar{x} - x\rangle \ge f(x)$. This implies that $v_f(\bar{x};x) \le 0$. Hence, $\omega_f(\bar{x}; v_f(\bar{x};x)) = 0$ and (3.2.5) holds.

Let $\langle g(x), x - \bar{x}\rangle > 0$. For

$$y = \tfrac{1}{\|g(x)\|}(\bar{x} + v_f(\bar{x};x) g(x))$$

we have $\langle g(x), y - \bar{x}\rangle = 0$ and $\| y - \bar{x} \| = v_f(\bar{x};x)$. Therefore

$$f(y) \ge f(x) + \langle g(x), y - x\rangle = f(x),$$

and

$$f(x) - f(\bar{x}) \le f(y) - f(\bar{x}) \le \omega_f(\bar{x}; \| y - \bar{x} \|) = \omega_f(\bar{x}; v_f(\bar{x};x)).$$

If f is Lipschitz continuous on $B_2(\bar{x},R)$ and $0 \le v_f(\bar{x};x) \le R$, then $y \in B_2(\bar{x},R)$. Hence,

$$f(x) - f(\bar{x}) \le f(y) - f(\bar{x}) \le M \| y - \bar{x} \| = M v_f(\bar{x};x).$$

□

Let us fix some x^*, a solution to problem (3.2.3). The values $v_f(x^*;x)$ allow us to estimate the quality of *localization sets.*

DEFINITION 3.2.1 *Let $\{x_i\}_{i=0}^{\infty}$ be a sequence in Q. Define*

$$S_k = \{x \in Q \mid \langle g(x_i), x_i - x\rangle \ge 0, \ i = 0 \dots k\}.$$

We call this set the localization set *of problem (3.2.3) generated by sequence* $\{x_i\}_{i=0}^{\infty}$.

Note that in view of inequality (3.2.4), for all $k \geq 0$ we have $x^* \in S_k$.
Denote

$$v_i = v_f(x^*; x_i) \ (\geq 0), \quad v_k^* = \min_{0 \leq i \leq k} v_i.$$

Thus,

$$v_k^* = \max\{r \mid \langle g(x_i), x_i - x \rangle \geq 0, \ i = 0 \ldots k, \ \forall x \in B_2(x^*, r)\}.$$

LEMMA 3.2.2 *Let* $f_k^* = \min_{0 \leq i \leq k} f(x_i)$. *Then* $f_k^* - f^* \leq \omega_f(x^*; v_k^*)$.

Proof: Using Lemma 3.2.1, we have

$$\omega_f(x^*; v_k^*) = \min_{0 \leq i \leq k} \omega_f(x^*; v_i) \geq \min_{0 \leq i \leq k} [f(x_i) - f^*] = f_k^* - f^*.$$

□

3.2.3 Subgradient method

Now we are ready to analyze the behavior of some minimization schemes. Consider the problem

$$\min\{f(x) \mid x \in Q\}, \tag{3.2.7}$$

where f is a convex on R^n function and Q is a *simple* closed convex set. The term "simple" means that we can solve *explicitly* some simple minimization problems over Q. In accordance to the goals of this section, we have to be able to find in a reasonably cheap way a Euclidean projection of any point onto Q.

We assume that problem (3.2.7) is equipped with a first-order oracle, which at any test point $\bar{x}$ provides us with the value of objective function $f(\bar{x})$ and with one of its subgradients $g(\bar{x})$.

As usual, we try first a version of a gradient method. Note that for nonsmooth problems the norm of the subgradient, $\| g(x) \|$, is not very informative. Therefore in the subgradient scheme we use a *normalized* direction $g(\bar{x})/ \parallel g(\bar{x}) \parallel$.

Subgradient method. Unconstrained minimization

0. Choose $x_0 \in Q$ and a sequence $\{h_k\}_{k=0}^{\infty}$:

$$h_k > 0, \quad h_k \to 0, \quad \sum_{k=0}^{\infty} h_k = \infty. \tag{3.2.8}$$

1. kth iteration ($k \geq 0$).

Compute $f(x_k)$, $g(x_k)$ and set

$$x_{k+1} = \pi_Q \left(x_k - h_k \frac{g(x_k)}{\|g(x_k)\|} \right).$$

Let us estimate the rate of convergence of this scheme.

THEOREM 3.2.2 *Let f be Lipschitz continuous on $B_2(x^*, R)$ with constant M and $x_0 \in B(x^*, R)$. Then*

$$f_k^* - f^* \leq M \frac{R^2 + \sum_{i=0}^{k} h_i^2}{2 \sum_{i=0}^{k} h_i}. \tag{3.2.9}$$

Proof: Denote $r_i = \parallel x_i - x^* \parallel$. Then, in view of Lemma 3.1.5, we have

$$\begin{aligned} r_{i+1}^2 &= \left\| \pi_Q \left(x_i - h_i \frac{g(x_i)}{\|g(x_i)\|} \right) - x^* \right\|^2 \\ &\leq \left\| x_i - h_i \frac{g(x_i)}{\|g(x_i)\|} - x^* \right\|^2 = r_i^2 - 2h_i v_i + h_i^2. \end{aligned}$$

Summing up these inequalities for $i = 0 \dots k$ we get

$$r_0^2 + \sum_{i=0}^{k} h_i^2 = 2 \sum_{i=0}^{k} h_i v_i + r_{k+1}^2 \geq 2 v_k^* \sum_{i=0}^{k} h_i.$$

Thus,

$$v_k^* \leq \frac{R^2 + \sum_{i=0}^{k} h_i^2}{2 \sum_{i=0}^{k} h_i}.$$

It remains to use Lemma 3.2.2. □

Thus, Theorem 3.2.2 demonstrates that the rate of convergence of *subgradient method* (3.2.8) depends on the values

$$\Delta_k = \frac{R^2 + \sum_{i=0}^{k} h_i^2}{2 \sum_{i=0}^{k} h_i}.$$

We can easily see that $\Delta_k \to 0$ if the series $\sum_{i=0}^{\infty} h_i$ diverges. However, let us try to choose h_k in an optimal way.

Let us assume that we have to perform a fixed number of steps of the subgradient method, say, N. Then, minimizing Δ_k as a function of $\{h_k\}_{k=0}^{N}$, we find that the optimal strategy is as follows:[2]

$$h_i = \frac{R}{\sqrt{N+1}}, \quad i = 0 \ldots N. \tag{3.2.10}$$

In this case $\Delta_N = \frac{R}{\sqrt{N+1}}$ and we obtain the following rate of convergence:

$$f_k^* - f^* \leq \frac{MR}{\sqrt{N+1}}.$$

Comparing this result with the lower bound of Theorem 3.2.1, we conclude:

> The subgradient method (3.2.8), (3.2.10) is optimal for problem (3.2.7) uniformly in the dimension n.

If we do not want to fix the number of iterations apriori, we can choose

$$h_i = \frac{r}{\sqrt{i+1}}, \quad i = 0, \ldots.$$

Then it is easy to see that Δ_k is proportional to

$$\frac{R^2 + r \ln(k+1)}{2r\sqrt{k+1}},$$

and we can classify the rate of convergence of this scheme as *sub-optimal*.

Thus, the simplest method for solving the problem (3.2.3) appears to be optimal. This indicates that the problems from our class are too complicated to be solved efficiently. However, we should remember, that our conclusion is valid *uniformly* in the dimension of the problem. We will see that a moderate dimension of the problem, taken into account in a proper way, helps to develop much more efficient schemes.

[2]From Example 3.1.2(3) we can see that Δ_k is a convex function of $\{h_i\}$.

3.2.4 Minimization with functional constraints

Let us apply a subgradient method to a constrained minimization problem with functional constraints. Consider the problem

$$\min\{f(x) \mid x \in Q,\ f_j(x) \le 0, i = 1 \dots m\}, \tag{3.2.11}$$

with convex f and f_j, and a simple bounded closed convex set Q:

$$\| x - y \| \le R, \quad \forall x, y \in Q.$$

Let us form an aggregate constraint $\bar{f}(x) = \left(\max\limits_{1 \le j \le m} f_j(x) \right)_+$. Then our problem can be written as follows:

$$\min\{f(x) \mid x \in Q,\ \bar{f}(x) \le 0\}. \tag{3.2.12}$$

Note that we can easily compute a subgradient $\bar{g}(x)$ of function $\bar{f}$, provided that we can do so for functions f_j (see Lemma 3.1.10).

Let us fix some x^*, a solution to (3.2.11). Note that $\bar{f}(x^*) = 0$ and $v_{\bar{f}}(x^*; x) \ge 0$ for all $x \in R^n$. Therefore, in view of Lemma 3.2.1 we have

$$\bar{f}(x) \le \omega_{\bar{f}}(x^*; v_{\bar{f}}(x^*; x)).$$

If f_j are Lipschitz continuous on Q with constant M, then for any x from R^n we have the estimate

$$\bar{f}(x) \le M \cdot v_{\bar{f}}(x^*; x).$$

Let us write down a subgradient minimization scheme for constrained minimization problem (3.2.12). We assume that R is known.

Subgradient method. Functional constraints

0. Choose $x_0 \in Q$ and sequence $\{h_k\}_{k=0}^{\infty}$:

$$h_k = \frac{R}{\sqrt{k+0.5}}.$$

1. kth iteration ($k \ge 0$). (3.2.13)

a). Compute $f(x_k)$, $g(x_k)$, $\bar{f}(x_k)$, $\bar{g}(x_k)$ and set

$$p_k = \begin{cases} g(x_k), & \text{if } \bar{f}(x_k) < \| \bar{g}(x_k) \| h_k, \quad (A), \\ \bar{g}(x_k), & \text{if } \bar{f}(x_k) \ge \| \bar{g}(x_k) \| h_k, \quad (B). \end{cases}$$

b). Set $x_{k+1} = \pi_Q \left(x_k - h_k \frac{p_k}{\| p_k \|} \right)$.

THEOREM 3.2.3 *Let f be Lipschitz continuous on $B_2(x^*, R)$ with constant M_1 and*

$$M_2 = \max_{1 \le j \le m} \{ \| g \| : \ g \in \partial f_j(x), \ x \in B_2(x^*, R) \}.$$

Then for any $k \ge 3$ there exists a number i', $0 \le i' \le k$, such that

$$f(x_{i'}) - f^* \le \frac{\sqrt{3} M_1 R}{\sqrt{k-1.5}}, \quad \bar{f}(x_{i'}) \le \frac{\sqrt{3} M_2 R}{\sqrt{k-1.5}}.$$

Proof: Note that for direction p_k, chosen in accordance to rule (B), we have

$$\| \bar{g}(x_k) \| \, h_k \le \bar{f}(x_k) \le \langle \bar{g}(x_k), x_k - x^* \rangle.$$

Hence, in this case $v_{\bar{f}}(x^*; x_k) \ge h_k$.

Let $k' = \left] \frac{k}{3} \right[$ and $I_k = \{ i \in [k', \ldots, k] : \ p_i = g(x_i) \}$. Denote

$$r_i = \| x_i - x^* \|, \quad v_i = v_f(x^*; x_i), \quad \bar{v}_i = v_{\bar{f}}(x^*; x_i).$$

Then for all i, $k' \le i \le k$, we have

$$\text{if} \quad i \in I_k, \quad \text{then} \quad r_{i+1}^2 \le r_i^2 - 2 h_i v_i + h_i^2,$$

$$\text{if} \quad i \notin I_k, \quad \text{then} \quad r_{i+1}^2 \le r_i^2 - 2 h_i \bar{v}_i + h_i^2.$$

Summing up these inequalities for $i \in [k', \ldots, k]$, we get:

$$r_{k'}^2 + \sum_{i=k'}^{k} h_i^2 \ge r_{k+1}^2 + 2 \sum_{i \in I_k} h_i v_i + 2 \sum_{i \notin I_k} h_i \bar{v}_i.$$

Recall that for $i \notin I_k$ we have $\bar{v}_i \ge h_i$ (Case (B)).

Assume that $v_i \ge h_i$ for all $i \in I_k$. Then

$$1 \ge \frac{1}{R^2} \sum_{i=k'}^{k} h_i^2 = \sum_{i=k'}^{k} \frac{1}{i+0.5} \ge \int_{k'}^{k+1} \frac{d\tau}{\tau+0.5} = \ln \frac{2k+3}{2k'+1} \ge \ln 3.$$

That is a contradiction. Thus, $I_k \ne \emptyset$ and there exists some $i' \in I_k$ such that $v_{i'} < h_{i'}$. Clearly, for this number we have $v_{i'} \le h_{k'}$, and, consequently, $(v_{i'})_+ \le h_{k'}$.

Thus, we conclude that $f(x_{i'}) - f^* \le M_1 h_{k'}$ (see Lemma 3.2.1) and, since $i' \in I_k$ we have also the estimate

$$\bar{f}(x_{i'}) \le \| \bar{g}(x_{i'}) \| \, h_{k'} \le M_2 h_{k'}.$$

It remains to note that $k' \ge \frac{k}{3} - 1$ and therefore $h_{k'} \le \frac{\sqrt{3} R}{\sqrt{k-1.5}}$. □

Comparing the result of Theorem 3.2.3 with the lower complexity bound of Theorem 3.2.1, we see that scheme (3.2.13) has an optimal rate of convergence. Recall, that this lower complexity bound was obtained for an unconstrained minimization problem. Thus, our result proves that from the viewpoint of analytical complexity the general convex unconstrained minimization problems are not easier than the constrained ones.

3.2.5 Complexity bounds in finite dimension

Let us look at the unconstrained minimization problem again, assuming that its dimension is relatively small. This means that our computational resources allow us to perform the number of iterations of a minimization method, proportional to the dimension of the space of variables. What will be the lower complexity bounds in this case?

In this section we obtain a finite-dimensional lower complexity bound for a problem, which is closely related to minimization problem. This is the *feasibility problem*:

$$\text{Find } x^* \in Q, \tag{3.2.14}$$

where Q is a convex set. We assume that this problem is endowed with an oracle, which answers our request at point $\bar{x} \in R^n$ in the following way:

- Either it reports that $\bar{x} \in Q$.
- Or, it returns a vector $\bar{g}$, separating $\bar{x}$ from Q:

$$\langle \bar{g}, \bar{x} - x \rangle \geq 0 \quad \forall x \in Q.$$

To estimate the complexity of this problem, we introduce the following assumption.

ASSUMPTION 3.2.1 *There exists a point $x^* \in Q$ such that for some $\epsilon > 0$ the ball $B_2(x^*, \epsilon)$ belongs to Q.*

For example, if we know an optimal value f^* for problem (3.2.3), we can treat this problem as a feasibility problem with

$$\bar{Q} = \{(t, x) \in R^{n+1} \mid t \geq f(x),\ t \leq f^* + \bar{\epsilon},\ x \in Q\}.$$

The relation between the accuracy parameters $\bar{\epsilon}$ and ϵ in (3.2.1) can be easily obtained, assuming that the function f is Lipschitz continuous. We leave this reasoning as an exercise for the reader.

Let us describe now a *resisting oracle* for problem (3.2.14). It forms a sequence of boxes $\{B_k\}_{k=0}^{\infty}$, $B_{k+1} \subset B_k$, defined by their lower and upper bounds.

$$B_k = \{x \in R^n \mid a_k \leq x \leq b_k\}.$$

For each box B_k, $k \geq 0$, denote by $c_k = \frac{1}{2}(a_k + b_k)$ its center. For boxes B_k, $k \geq 1$, the oracle creates an individual separating vector g_k. Up to a sign change, this is always a coordinate vector.

In the scheme below we use two dynamic counters:

- m is the number of generated boxes.
- i is the active coordinate.

Denote by $e \in R^n$ a vector of all 1s. The oracle starts from the following settings:

$$a_0 := -Re, \quad b_0 := Re, \quad m := 0, \quad i := 1.$$

Its input is an arbitrary $x \in R^n$.

Resisting oracle. Feasibility problem

If $x \notin B_0$ **then** return a separator of x from B_0 **else**

1. Find the maximal $k \in [0, \dots, m]$: $x \in B_k$.

2. If $k < m$ **then** return g_k **else** {Create a new box}:

If $x^{(i)} \geq c_m^{(i)}$ **then** $a_{m+1} := a_m$,

$$b_{m+1} := b_m + (c_m^{(i)} - b_m^{(i)})e_i, \quad g_m := e_i.$$

else $a_{m+1} := a_m + (c_m^{(i)} - a_m^{(i)})e_i$,

$$b_{m+1} := b_m, \quad g_m := -e_i.$$

$m := m + 1$; $i := i + 1$; **If** $i > n$ **then** $i := 1$.

Return g_m.

This oracle implements a very simple strategy. Note, that the next box B_{m+1} is always a half of the last box B_m. The box B_m is divided into two parts by a hyperplane, which passes through its center and which corresponds to the active coordinate i. Depending in which part of the box B_m we get the test point x, we choose the sign of the separation vector $g_{m+1} = \pm e_i$. After creating a new box B_{m+1} the index i is increased by 1. If this value exceeds n, we return again to $i = 1$. Thus, the sequence of boxes $\{B_k\}$ possesses two important properties:

- $\mathrm{vol}_n B_{k+1} = \frac{1}{2}\mathrm{vol}_n B_k$.
- For any $k \geq 0$ we have $b_{k+n} - a_{k+n} = \frac{1}{2}(b_k - a_k)$.

Note also that the number of generated boxes does not exceed the number of calls of the oracle.

LEMMA 3.2.3 *For all $k \geq 0$ we have the inclusion*

$$B_2(c_k, r_k) \subset B_k, \quad \text{with} \quad r_k = \frac{R}{2}\left(\frac{1}{2}\right)^{-\frac{k}{n}}. \tag{3.2.15}$$

Proof: Indeed, for all $k \in [0, \ldots, n-1]$ we have

$$B_k \supset B_n = \{x \mid c_n - \tfrac{1}{2}Re \leq x \leq c_n + \tfrac{1}{2}Re\} \supset B_2(c_n, \tfrac{1}{2}R).$$

Therefore, for such k we have $B_k \supset B_2(c_k, \frac{1}{2}R)$ and (3.2.15) holds. Further, let $k = nl + p$ with some $p \in [0, \ldots, n-1]$. Since

$$b_k - a_k = \left(\tfrac{1}{2}\right)^{-l}(b_p - a_p),$$

we conclude that

$$B_k \supset B_2\left(c_k, \tfrac{1}{2}R\left(\tfrac{1}{2}\right)^{-l}\right).$$

It remains to note that $r_k \leq \frac{1}{2}R\left(\frac{1}{2}\right)^{-l}$. □

Lemma 3.2.3 immediately leads to the following complexity result.

THEOREM 3.2.4 *Consider a class of feasibility problems (3.2.14), which satisfy Assumption 3.2.1, and for which the feasible sets Q belong to $B_\infty(0, R)$. The lower analytical complexity bound for this class is $n \ln \frac{R}{2\epsilon}$ calls of the oracle.*

Proof: Indeed, we have seen that the number of generated boxes does not exceed the number of calls of the oracle. Moreover, in view of Lemma 3.2.3, after k iterations the last box contains the ball $B_2(c_{m_k}, r_k)$. □

The lower complexity bound for minimization problem (3.2.3) can be obtained in a similar way. However, the corresponding reasoning is more complicated. Therefore we present here only a conclusion.

THEOREM 3.2.5 *The lower bound for analytical complexity of problem class formed by minimization problems (3.2.3) with $Q \subseteq B_\infty(0,R)$ and $f \in \mathcal{F}_M^{0,0}(B_\infty(0,R))$, is $n \ln \frac{MR}{8\epsilon}$ calls of the oracle.* □

3.2.6 Cutting plane schemes

Let us look now at the following constrained minimization problem:

$$\min\{f(x) \mid x \in Q\}, \tag{3.2.16}$$

where f is a function convex on R^n, and Q is a bounded closed convex set such that

$$\operatorname{int} Q \neq \emptyset, \quad \operatorname{diam} Q = D < \infty.$$

We assume that Q is not simple and that our problem is equipped with a separating oracle. At any test point $\bar{x} \in R^n$ this oracle returns a vector g which is:

- a subgradient of f at $\bar{x}$, if $x \in Q$,
- a separator of $\bar{x}$ from Q, if $x \notin Q$.

An important example of such a problem is a constrained minimization problem with functional constraints (3.2.11). We have seen that this problem can be rewritten as a problem with a single functional constraint (see (3.2.12)), which defines a feasible set

$$Q = \{x \in R^n \mid \bar{f}(x) \leq 0\}.$$

In this case, for $x \notin Q$ the oracle has to provide us with any subgradient $\bar{g} \in \partial \bar{f}(x)$. Clearly, $\bar{g}$ separates x from Q (see Theorem 3.1.16).

Let us present the main property of finite-dimensional localization sets.

Consider a sequence $X \equiv \{x_i\}_{i=0}^{\infty}$ belonging to the set Q. Recall, that the localization sets, generated by this sequence, are defined as follows:

$$\begin{aligned} S_0(X) &= Q, \\ S_{k+1}(X) &= \{x \in S_k(X) \mid \langle g(x_k), x_k - x\rangle \geq 0\}. \end{aligned}$$

Clearly, for any $k \geq 0$ we have $x^* \in S_k$. Denote

$$v_i = v_f(x^*; x_i) \; (\geq 0), \quad v_k^* = \min_{0 \leq i \leq k} v_i.$$

Denote by $\mathrm{vol}_n S$ an n-dimensional volume of set $S \subset R^n$.

THEOREM 3.2.6 *For any $k \geq 0$ we have*

$$v_k^* \leq D \left[\frac{\mathrm{vol}_n S_k(X)}{\mathrm{vol}_n Q} \right]^{\frac{1}{n}}.$$

Proof: Denote $\alpha = v_k^*/D$ (≤ 1). Since $Q \subseteq B_2(x^*, D)$ we have the following inclusion:

$$(1-\alpha)x^* + \alpha Q \subseteq (1-\alpha)x^* + \alpha B_2(x^*, D) = B_2(x^*, v_k^*).$$

Since Q is convex, we conclude that

$$(1-\alpha)x^* + \alpha Q \equiv [(1-\alpha)x^* + \alpha Q] \bigcap Q \subseteq B_2(x^*, v_k^*) \bigcap Q \subseteq S_k(X).$$

Therefore $\mathrm{vol}_n S_k(X) \geq \mathrm{vol}_n [(1-\alpha)x^* + \alpha Q] = \alpha^n \mathrm{vol}_n Q$. □

Quite often the set Q is rather complicated and it is difficult to work directly with sets $S_k(X)$. Instead, we can update some simple *upper* approximations of these sets. The process of generating such approximations is described by the following *cutting plane* scheme.

General cutting plane scheme

0. Choose a bounded set $E_0 \supseteq Q$.

1. kth iteration $(k \geq 0)$.

a) Choose $y_k \in E_k$

b) If $y_k \in Q$ then compute $f(y_k)$, $g(y_k)$. If $y_k \notin Q$, then compute $\bar{g}(y_k)$, which separates y_k from Q. (3.2.17)

c) Set

$$g_k = \begin{cases} g(y_k), & \text{if } y_k \in Q, \\ \bar{g}(y_k), & \text{if } y_k \notin Q. \end{cases}$$

d) Choose $E_{k+1} \supseteq \{x \in E_k \mid \langle g_k, y_k - x \rangle \geq 0\}$.

Let us estimate the performance of the above process. Consider the sequence $Y = \{y_k\}_{k=0}^{\infty}$, involved in this scheme. Denote by X a subsequence of feasible points in the sequence Y: $X = Y \bigcap Q$. Let us introduce the counter

$$i(k) = \text{number of points } y_j,\ 0 \leq j < k, \text{ such that } y_j \in Q.$$

Thus, if $i(k) > 0$, then $X \neq \emptyset$.

LEMMA 3.2.4 *For any $k \geq 0$, we have $S_{i(k)} \subseteq E_k$.*

Proof: Indeed, if $i(0) = 0$, then $S_0 = Q \subseteq E_0$. Let us assume that $S_{i(k)} \subseteq E_k$ for some $k \geq 0$. Then, at the next iteration there are two possibilities:

a) $i(k+1) = i(k)$. This happens if and only if $y_k \notin Q$. Then

$$\begin{aligned} E_{k+1} &\supseteq \{x \in E_k \mid \langle \bar{g}(y_k), y_k - x\rangle \geq 0\} \\ &\supseteq \{x \in S_{i(k+1)} \mid \langle \bar{g}(y_k), y_k - x\rangle \geq 0\} = S_{i(k+1)} \end{aligned}$$

since $S_{i(k+1)} \subseteq Q$ and $\bar{g}(y_k)$ separates y_k from Q.

b) $i(k+1) = i(k) + 1$. In this case $y_k \in Q$. Then

$$\begin{aligned} E_{k+1} &\supseteq \{x \in E_k \mid \langle g(y_k), y_k - x\rangle \geq 0\} \\ &\supseteq \{x \in S_{i(k)} \mid \langle g(y_k), y_k - x\rangle \geq 0\} = S_{i(k)+1} \end{aligned}$$

since $y_k = x_{i(k)}$. □

The above results immediately lead to the following important conclusion.

COROLLARY 3.2.1 *1. For any k such that $i(k) > 0$, we have*

$$v^*_{i(k)}(X) \leq D\left[\frac{\mathrm{vol}_n S_{i(k)}(X)}{\mathrm{vol}_n Q}\right]^{\frac{1}{n}} \leq D\left[\frac{\mathrm{vol}_n E_k}{\mathrm{vol}_n Q}\right]^{\frac{1}{n}}.$$

2. If $\mathrm{vol}_n E_k < \mathrm{vol}_n Q$, then $i(k) > 0$.

Proof: We have already proved the first statement. The second one follows from the inclusion $Q = S_0 = S_{i(k)} \subseteq E_k$, which is valid for all k such that $i(k) = 0$. □

Thus, if we manage to ensure $\mathrm{vol}_n E_k \to 0$, then we obtain a convergent scheme. Moreover, the rate of decrease of the volume automatically defines the rate of convergence of the method. Clearly, we should try to decrease $\mathrm{vol}_n E_k$ as fast as possible.

Historically, the first nonsmooth minimization method, implementing the idea of cutting planes, was the *center of gravity* method. It is based on the following geometrical fact.

Consider a bounded convex set $S \subset R^n$, $\mathrm{int}\, S \neq \emptyset$. Define the *center of gravity* of this set as

$$cg(S) = \frac{1}{\mathrm{vol}_n S} \int_S x dx.$$

The following result demonstrates that any cut passing through the center of gravity divides the set on two proportional pieces.

LEMMA 3.2.5 *Let g be a direction in R^n. Define*

$$S_+ = \{x \in S \mid \langle g, cg(S) - x\rangle \geq 0\}.$$

Then

$$\frac{\mathrm{vol}_n S_+}{\mathrm{vol}_n S} \leq 1 - \frac{1}{e}.$$

(We accept this result without proof.) □

This observation naturally leads to the following minimization scheme.

Method of centers of gravity
0. Set $S_0 = Q$. **1.** kth iteration ($k \geq 0$). a) Choose $x_k = cg(S_k)$ and compute $f(x_k)$, $g(x_k)$. b) Set $S_{k+1} = \{x \in S_k \mid \langle g(x_k), x_k - x\rangle \geq 0\}$.

Let us estimate the rate of convergence of this method. Denote

$$f_k^* = \min_{0 \leq j \leq k} f(x_j).$$

THEOREM 3.2.7 *If f is Lipschitz continuous on $B_2(x^*, D)$ with a constant M, then for any $k \geq 0$ we have*

$$f_k^* - f^* \leq MD\left(1 - \frac{1}{e}\right)^{-\frac{k}{n}}.$$

Proof: The statement follows from Lemma 3.2.2, Theorem 3.2.6 and Lemma 3.2.5. □

Comparing this result with the lower complexity bound of Theorem 3.2.5, we see that the center-of-gravity method is optimal in finite dimension. Its rate of convergence does not depend on any individual characteristics of our problem like condition number, etc. However, we should accept that this method is absolutely impractical, since the computation of the center of gravity in multi-dimensional space is a more difficult problem than our initial one.

Let us look at another method, which uses a possibility of approximation of the localization sets. This method is based on the following geometrical observation.

Let H be a positive definite symmetric $n \times n$ matrix. Consider the *ellipsoid*

$$E(H, \bar{x}) = \{x \in R^n \mid \langle H^{-1}(x - \bar{x}), x - \bar{x}\rangle \leq 1\}.$$

Let us choose a direction $g \in R^n$ and consider a half of the above ellipsoid, defined by corresponding hyperplane:

$$E_+ = \{x \in E(H, \bar{x}) \mid \langle g, \bar{x} - x\rangle \geq 0\}.$$

It turns out that this set belongs to another ellipsoid, which volume is strictly smaller than the volume of $E(H, \bar{x})$.

LEMMA 3.2.6 *Denote*

$$\begin{aligned} \bar{x}_+ &= \bar{x} - \tfrac{1}{n+1} \cdot \tfrac{Hg}{\langle Hg, g\rangle^{1/2}}, \\ H_+ &= \tfrac{n^2}{n^2-1}\left(H - \tfrac{2}{n+1} \cdot \tfrac{Hgg^T H}{\langle Hg, g\rangle}\right). \end{aligned}$$

Then $E_+ \subset E(H_+, \bar{x}_+)$ *and*

$$\text{vol}_n\, E(H_+, \bar{x}_+) \leq \left(1 - \tfrac{1}{(n+1)^2}\right)^{\frac{n}{2}} \text{vol}_n\, E(H, \bar{x}).$$

Proof: Denote $G = H^{-1}$ and $G_+ = H_+^{-1}$. It is clear that

$$G_+ = \tfrac{n^2-1}{n^2}\left(G + \tfrac{2}{n-1} \cdot \tfrac{gg^T}{\langle Hg, g\rangle}\right).$$

Without loss of generality we can assume that $\bar{x} = 0$ and $\langle Hg, g\rangle = 1$. Suppose $x \in E_+$. Note that $\bar{x}_+ = -\frac{1}{n+1}Hg$. Therefore

$$\begin{aligned} \| x - \bar{x}_+ \|^2_{G_+} &= \tfrac{n^2-1}{n^2}\left(\| x - \bar{x}_+ \|^2_G + \tfrac{2}{n-1}\langle g, x - \bar{x}_+\rangle^2\right), \\ \| x - \bar{x}_+ \|^2_G &= \| x \|^2_G + \tfrac{2}{n+1}\langle g, x\rangle + \tfrac{1}{(n+1)^2}, \\ \langle g, x - \bar{x}_+\rangle^2 &= \langle g, x\rangle^2 + \tfrac{2}{n+1}\langle g, x\rangle + \tfrac{1}{(n+1)^2}. \end{aligned}$$

Putting all terms together, we obtain

$$\| x - \bar{x}_+ \|^2_{G_+} = \tfrac{n^2-1}{n^2}\left(\| x \|^2_G + \tfrac{2}{n-1}\langle g, x\rangle^2 + \tfrac{2}{n-1}\langle g, x\rangle + \tfrac{1}{n^2-1}\right).$$

Note that $\langle g, x\rangle \leq 0$ and $\| x \|_G \leq 1$. Therefore

$$\langle g, x\rangle^2 + \langle g, x\rangle = \langle g, x\rangle(1 + \langle g, x\rangle) \leq 0.$$

Hence,

$$\| x - \bar{x}_+ \|^2_{G_+} \leq \frac{n^2-1}{n^2} \left(\| x \|^2_G + \frac{1}{n^2-1} \right) \leq 1.$$

Thus, we have proved that $E_+ \subset E(H_+, \bar{x}_+)$.

Let us estimate the volume of $E(H_+, \bar{x}_+)$.

$$\begin{aligned} \frac{\text{vol}_n E(H_+,\bar{x}_+)}{\text{vol}_n E(H,\bar{x})} &= \left[\frac{\det H_+}{\det H} \right]^{1/2} = \left[\left(\frac{n^2}{n^2-1} \right)^n \frac{n-1}{n+1} \right]^{1/2} \\ &= \left[\frac{n^2}{n^2-1} \left(1 - \frac{2}{n+1} \right)^{\frac{1}{n}} \right]^{\frac{n}{2}} \leq \left[\frac{n^2}{n^2-1} \left(1 - \frac{2}{n(n+1)} \right) \right]^{\frac{n}{2}} \\ &= \left[\frac{n^2(n^2+n-2)}{n(n-1)(n+1)^2} \right]^{\frac{n}{2}} = \left[1 - \frac{1}{(n+1)^2} \right]^{\frac{n}{2}}. \end{aligned}$$

□

It turns out that the ellipsoid $E(H_+, \bar{x}_+)$ is the ellipsoid of the *minimal* volume containing the half of the initial ellipsoid E_+.

Our observations can be implemented in the following algorithmic scheme of the *ellipsoid method*.

Ellipsoid method
0. Choose $y_0 \in R^n$ and $R > 0$ such that $B_2(y_0, R) \supseteq Q$. Set $H_0 = R^2 \cdot I_n$.   **1.** kth iteration ($k \geq 0$). $$g_k = \begin{cases} g(y_k), & \text{if } y_k \in Q, \\ \bar{g}(y_k), & \text{if } y_k \notin Q, \end{cases}$$ $$y_{k+1} = y_k - \frac{1}{n+1} \cdot \frac{H_k g_k}{\langle H_k g_k, g_k \rangle^{1/2}},$$ $$H_{k+1} = \frac{n^2}{n^2-1} \left(H_k - \frac{2}{n+1} \cdot \frac{H_k g_k g_k^T H_k}{\langle H_k g_k, g_k \rangle} \right).$$

(3.2.18)

This method can be seen as a particular implementation of general scheme (3.2.17) by choosing

$$E_k = \{ x \in R^n \mid \langle H_k^{-1}(x - y_k), x - y_k \rangle \leq 1 \}$$

and y_k being the center of this ellipsoid.

Let us present an efficiency estimate for the ellipsoid method. Denote $Y = \{y_k\}_{k=0}^{\infty}$ and let X be a feasible subsequence of the sequence Y:

$$X = Y \bigcap Q.$$

Denote $f_k^* = \min_{0 \leq j \leq k} f(x_j)$.

THEOREM 3.2.8 *Let f be Lipschitz continuous on $B_2(x^*, R)$ with some constant M. Then for $i(k) > 0$, we have*

$$f_{i(k)}^* - f^* \leq MR\left(1 - \frac{1}{(n+1)^2}\right)^{\frac{k}{2}} \cdot \left[\frac{\text{vol}_n B_0(x_0,R)}{\text{vol}_n Q}\right]^{\frac{1}{n}}.$$

Proof: The proof follows from Lemma 3.2.2, Corollary 3.2.1 and Lemma 3.2.6. □

We need additional assumptions to guarantee $X \neq \emptyset$. Assume that there exists some $\rho > 0$ and $\bar{x} \in Q$ such that

$$B_2(\bar{x}, \rho) \subseteq Q. \tag{3.2.19}$$

Then

$$\left[\frac{\text{vol}_n E_k}{\text{vol}_n Q}\right]^{\frac{1}{n}} \leq \left(1 - \frac{1}{(n+1)^2}\right)^{\frac{k}{2}} \left[\frac{\text{vol}_n B_2(x_0,R)}{\text{vol}_n Q}\right]^{\frac{1}{n}} \leq \frac{1}{\rho} e^{-\frac{k}{2(n+1)^2}} R.$$

In view of Corollary 3.2.1, this implies that $i(k) > 0$ for all

$$k > 2(n+1)^2 \ln \frac{R}{\rho}.$$

If $i(k) > 0$, then

$$f_{i(k)}^* - f^* \leq \frac{1}{\rho} MR^2 \cdot e^{-\frac{k}{2(n+1)^2}}.$$

In order to ensure that(3.2.19) holds for a constrained minimization problem with functional constraints, it is enough to assume that all constraints are Lipschitz continuous and there is a feasible point, at which all functional constraints are *strictly negative* (Slater condition). We leave the details of the proof as an exercise for the reader.

Let us discuss now the complexity of ellipsoid method (3.2.18). Each iteration of this method is rather cheap; it takes only $O(n^2)$ arithmetic operations. On the other hand, in order to generate an ϵ-solution of problem (3.2.16), satisfying assumption (3.2.19), this method needs

$$2(n+1)^2 \ln \frac{MR^2}{\rho\epsilon}$$

calls of the oracle. This efficiency estimate is not optimal (see Theorem 3.2.5), but it has a polynomial dependence on $\ln \frac{1}{\epsilon}$ and a polynomial dependence on logarithms of the class parameters M, R and ρ. For problem classes, whose oracle has a polynomial complexity, such algorithms are called (weakly) *polynomial.*

To conclude this section, let us mention that there are several methods that work with localization sets in the form of a polytope:

$$E_k = \{x \in R^n \mid \langle a_j, x\rangle \leq b_j, \ j = 1 \dots m_k\}.$$

Let us list the most important methods of this type:

- *Inscribed Ellipsoid Method.* The point y_k in this scheme is chosen as follows:

$$y_k = \text{Center of the maximal ellipsoid } W_k : \ W_k \subset E_k.$$

- *Analytic Center Method.* In this method the point y_k is chosen as the minimum of the *analytic barrier*

$$F_k(x) = -\sum_{j=1}^{m_k} \ln(b_j - \langle a_j, x\rangle).$$

- *Volumetric Center Method.* This is also a barrier-type scheme. The point y_k is chosen as the minimum of the *volumetric barrier*

$$V_k(x) = \ln \det F_k''(x),$$

where $F_k(x)$ is the analytic barrier for the set E_k.

All these methods are polynomial with complexity bound

$$n\left(\ln \tfrac{1}{\epsilon}\right)^p,$$

where p is either 1 or 2. However, the complexity of each iteration in these methods is much larger ($n^3 \div n^4$ arithmetic operations). In the next chapter we will see that the test point y_k for these schemes can be computed by *interior-point* methods.

3.3 Methods with complete data

(Model of nonsmooth function; Kelley method; Idea of level method; Unconstrained minimization; Efficiency estimates; Problems with functional constraints.)

3.3.1 Model of nonsmooth function

In the previous section we studied several methods for solving the following problem:

$$\min_{x \in Q} f(x), \tag{3.3.1}$$

where f is a Lipschitz continuous convex function and Q is a closed convex set. We have seen that the optimal method for problem (3.3.1) is the *subgradient method* (3.2.8), (3.2.10). Note, that this conclusion is valid for the *whole* class of Lipschitz continuous functions. However, when we are going to minimize a particular function from that class, we can expect that it is not too bad. We can hope that the real performance of the minimization method will be much better than a theoretical bound derived from a worst-case analysis. Unfortunately, as far as the subgradient method is concerned, these expectations are too optimistic. The scheme of the subgradient method is very strict and in general it *cannot* converge faster than in theory. It can be also shown that the ellipsoid method (3.2.18), inherits this drawback of subgradient scheme. In practice it works more or less in accordance to its theoretical bound even when it is applied to a very simple function like $\| x \|^2$.

In this section we will discuss the algorithmic schemes, which are more flexible than the subgradient and the ellipsoid methods. These schemes are based on the notion of a *model* of nonsmooth function.

DEFINITION 3.3.1 *Let $X = \{x_k\}_{k=0}^{\infty}$ be a sequence in Q. Denote*

$$\hat{f}_k(X; x) = \max_{0 \le i \le k} [f(x_i) + \langle g(x_i), x - x_i \rangle],$$

where $g(x_i)$ are some subgradients of f at x_i.

The function $\hat{f}_k(X; x)$ is called the model *of convex function f.*

Note that $f_k(X; x)$ is a piece-wise linear function of x. In view of inequality (3.1.10) we always have

$$f(x) \ge \hat{f}_k(X; x)$$

for all $x \in R^n$. However, at all test points x_i, $0 \le i \le k$, we have

$$f(x_i) = \hat{f}_k(X; x_i), \quad g(x_i) \in \partial \hat{f}_k(X; x_i).$$

The next model is always better than the previous one:

$$\hat{f}_{k+1}(X; x) \ge \hat{f}_k(X; x)$$

for all $x \in R^n$.

3.3.2 Kelley method

Model $\hat{f}_k(X;x)$ represents the *complete* information on function f accumulated after k calls of the oracle. Therefore it seems natural to develop a minimization scheme, based on this object. Perhaps the most natural method of this type is as follows:

Kelley method
0. Choose $x_0 \in Q$.   **1.** kth iteration ($k \geq 0$).   Find $x_{k+1} \in \operatorname{Arg}\min\limits_{x \in Q} \hat{f}_k(X;x)$.

(3.3.2)

Intuitively, this scheme looks very attractive. Even the presence of a complicated auxiliary problem is not too disturbing, since it can be solved by linear optimization methods in finite time. However, it turns out that this method cannot be recommended for practical applications. And the main reason for that is its instability. Note that the solution of auxiliary problem in the method (3.3.2) may be not unique. Moreover, the whole set $\operatorname{Arg}\min\limits_{x \in Q} \hat{f}_k(X;x)$ can be unstable with respect to an arbitrary small variation of data $\{f(x_i), g(x_i)\}$. This feature results in unstable practical behavior of the scheme. Moreover, this feature can be used for constructing an example of a problem, in which the Kelley method has a very bad *lower* complexity bound.

EXAMPLE 3.3.1 Consider the problem (3.3.1) with

$$
\begin{aligned}
f(y,x) &= \max\{\mid y \mid, \parallel x \parallel^2\}, \quad y \in R^1, \ x \in R^n, \\
Q &= \{z = (y,x) : \ y^2 + \parallel x \parallel^2 \leq 1\}.
\end{aligned}
$$

Thus, the solution of this problem is $z^* = (y^*, x^*) = (0,0)$, and the optimal value $f^* = 0$. Denote by $Z_k^* = \operatorname{Arg}\min\limits_{z \in Q} \hat{f}_k(Z;z)$, the optimal set of model $\hat{f}_k(Z;z)$, and by $\hat{f}_k^* = \hat{f}_k(Z_k^*)$ the optimal value of the model.

Let us choose $z_0 = (1,0)$. Then the initial model of function f is $\hat{f}_0(Z;z) = y$. Therefore, the first point, generated by the Kelley method is $z_1 = (-1,0)$. Hence, the next model of the function f is as follows:

$$
\hat{f}_1(Z;z) = \max\{y, -y\} = \mid y \mid .
$$

Clearly, $\hat{f}_1^* = 0$. Note that $\hat{f}_{k+1}^* \geq \hat{f}_k^*$. On the other hand,

$$\hat{f}_k^* \leq f(z^*) = 0.$$

Thus, for all consequent models with $k \geq 1$ we will have $\hat{f}_k^* = 0$ and $Z_k^* = (0, X_k^*)$, where

$$X_k^* = \{x \in B_2(0,1) : \| x_i \|^2 + \langle 2x_i, x - x_i \rangle \leq 0, \; i = 0 \dots k\}.$$

Let us estimate efficiency of the cuts for the set X_k^*. Since x_{k+1} can be an *arbitrary* point from X_k^*, at the first stage of the method we can choose x_i with the unit norms: $\| x_i \| = 1$. Then the set X_k^* is defined as follows:

$$X_k^* = \{x \in B_2(0,1) \mid \langle x_i, x \rangle \leq \tfrac{1}{2}, i = 0 \dots k\}.$$

We can do that if

$$S_2(0,1) \equiv \{x \in R^n \mid \| x \| = 1\} \bigcap X_k^* \neq \emptyset.$$

As far as this is possible, we can have

$$f(z_i) \equiv f(0, x_i) = 1.$$

Let us estimate a possible length of this stage using the following fact.

> Let d be a direction in R^n, $\| d \| = 1$. Consider a surface
>
> $$S(\alpha) = \{x \in R^n \mid \| x \| = 1, \; \langle d, x \rangle \geq \alpha\}, \quad \alpha \in [\tfrac{1}{2}, 1].$$
>
> Then $v(\alpha) \equiv \mathrm{vol}_{n-1}(S(\alpha)) \leq v(0) \left[1 - \alpha^2\right]^{\frac{n-1}{2}}$.

At the first stage, each step cuts from the sphere $S_2(0,1)$ at most the segment $S(\frac{1}{2})$. Therefore, we can continue the process if $k \leq \left[\frac{2}{\sqrt{3}}\right]^{n-1}$. During all these iterations we still have $f(z_i) = 1$.

Since at the first stage of the process the cuts are $\langle x_i, x \rangle \leq \frac{1}{2}$, for all k, $0 \leq k \leq N \equiv \left[\frac{2}{\sqrt{3}}\right]^{n-1}$, we have

$$B_2(0, \tfrac{1}{2}) \subset X_k^*.$$

This means that after N iterations we can repeat our process with the ball $B_2(0, \frac{1}{2})$, etc. Note that $f(0, x) = \frac{1}{4}$ for all x from $B_2(0, \frac{1}{2})$.

Thus, we prove the following *lower* bound for the Kelley method (3.3.2):

$$f(x_k) - f^* \geq \left(\frac{1}{4}\right)^{k\left[\frac{\sqrt{3}}{2}\right]^{n-1}}.$$

This means that we cannot get an ϵ-solution of our problem less than in

$$\frac{\ln \frac{1}{\epsilon}}{2 \ln 2} \left[\frac{2}{\sqrt{3}}\right]^{n-1}$$

calls of the oracle. It remains to compare this lower bound with the upper complexity bounds of other methods:

Ellipsoid method:	$O\left(n^2 \ln \frac{1}{\epsilon}\right)$
Optimal methods:	$O\left(n \ln \frac{1}{\epsilon}\right)$
Gradient method:	$O\left(\frac{1}{\epsilon^2}\right)$

□

3.3.3 Level method

Let us show that it is possible to work with models in a stable way. Denote

$$\hat{f}_k^* = \min_{x \in Q} \hat{f}_k(X; x), \quad f_k^* = \min_{0 \leq i \leq k} f(x_i).$$

The first value is called the *minimal value* of the model, and the second one the *record value* of the model. Clearly $\hat{f}_k^* \leq f^* \leq f_k^*$.

Let us choose some $\alpha \in (0, 1)$. Denote

$$l_k(\alpha) = (1 - \alpha)\hat{f}_k^* + \alpha f_k^*.$$

Consider the level set

$$\mathcal{L}_k(\alpha) = \{x \in Q \mid f_k(x) \leq l_k(\alpha)\}.$$

Clearly, $\mathcal{L}_k(\alpha)$ is a closed convex set.

Note that the set $\mathcal{L}_k(\alpha)$ is of a certain interest for an optimization scheme. Firstly, inside this set clearly there are no test points of the current model. Secondly, this set is stable with respect to a small variation of the data. Let us present a minimization scheme, which deals directly with this level set.

Level method

0. Choose point $x_0 \in Q$, accuracy $\epsilon > 0$ and level coefficient $\alpha \in (0,1)$.
1. kth iteration ($k \geq 0$).
   a). Compute $\hat{f}_k^*$ and f_k^*.
   b). If $f_k^* - \hat{f}_k^* \leq \epsilon$, then STOP.
   c). Set $x_{k+1} = \pi_{\mathcal{L}_k(\alpha)}(x_k)$.

(3.3.3)

In this scheme there are two quite expensive operations. We need to compute an optimal value $\hat{f}_k^*$ of the current model. If Q is a polytope, then this value can be obtained from the following linear programming problem:

$$\begin{aligned} &\min \quad t, \\ &\text{s.t.} \quad f(x_i) + \langle g(x_i), x - x_i \rangle \leq t, \; i = 0 \ldots k, \\ &x \in Q. \end{aligned}$$

We also need to compute projection $\pi_{\mathcal{L}_k(\alpha)}(x_k)$. If Q is a polytope, then this is a quadratic programming problem:

$$\begin{aligned} &\min \quad \| x - x_k \|^2, \\ &\text{s.t.} \quad f(x_i) + \langle g(x_i), x - x_i \rangle \leq l_k(\alpha), \; i = 0 \ldots k, \\ &x \in Q. \end{aligned}$$

Both these problems are solvable either by a standard simplex-type method, or by interior-point schemes.

Let us look at some properties of the level method. Recall, that the optimal values of the model decrease, and the records values increase:

$$\hat{f}_k^* \leq \hat{f}_{k+1}^* \leq f^* \leq f_{k+1}^* \leq f_k^*.$$

Denote $\Delta_k = [\hat{f}_k^*, f_k^*]$ and $\delta_k = f_k^* - \hat{f}_k^*$. We call δ_k the *gap* of the model $\hat{f}_k(X; x)$. Then

$$\Delta_{k+1} \subseteq \Delta_k, \quad \delta_{k+1} \leq \delta_k.$$

The next result is crucial for the analysis of the level method.

LEMMA 3.3.1 *Assume that for some $p \geq k$ we have $\delta_p \geq (1-\alpha)\delta_k$. Then for all i, $k \leq i \leq p$,*

$$l_i(\alpha) \geq \hat{f}_p^*$$

Proof: Note that for such i we have $\delta_p \geq (1-\alpha)\delta_k \geq (1-\alpha)\delta_i$. Therefore

$$l_i(\alpha) = f_i^* - (1-\alpha)\delta_i \geq f_p^* - (1-\alpha)\delta_i = \hat{f}_p^* + \delta_p - (1-\alpha)\delta_i \geq \hat{f}_p^*.$$

□

Let us show that the steps of the level method are large enough. Denote

$$M_f = \max\{\| g \| \mid g \in \partial f(x),\ x \in Q\}.$$

LEMMA 3.3.2 *For the sequence $\{x_k\}$ generated by the level method we have*

$$\| x_{k+1} - x_k \| \geq \frac{(1-\alpha)\delta_k}{M_f}.$$

Proof: Indeed,

$$\begin{aligned} f(x_k) - (1-\alpha)\delta_k &\geq f_k^* - (1-\alpha)\delta_k = l_k(\alpha) \\ &\geq \hat{f}_k(x_{k+1}) \geq f(x_k) + \langle g(x_k), x_{k+1} - x_k \rangle \\ &\geq f(x_k) - M_f \| x_{k+1} - x_k \| . \end{aligned}$$

□

Finally, we need to show that the gap in the model is decreasing.

LEMMA 3.3.3 *Let Q in the problem (3.3.1) be bounded:* $\operatorname{diam} Q \leq D$. *If for some $p \geq k$ we have $\delta_p \geq (1-\alpha)\delta_k$, then*

$$p + 1 - k \leq \frac{M_f^2 D^2}{(1-\alpha)^2 \delta_p^2}.$$

Proof: Denote $x_k^* \in \operatorname{Arg}\min\limits_{x \in Q} \hat{f}_k(X; x)$. In view of Lemma 3.3.1 we have

$$\hat{f}_i(X; x_p^*) \leq \hat{f}_p(X; x_p^*) = \hat{f}_p^* \leq l_i(\alpha)$$

for all i, $k \leq i \leq p$. Therefore, in view of Lemma 3.1.5 and Lemma 3.3.2 we obtain the following:

$$\begin{aligned} \| x_{i+1} - x_p^* \|^2 &\leq \| x_i - x_p^* \|^2 - \| x_{i+1} - x_i \|^2 \\ &\leq \| x_i - x_p^* \|^2 - \frac{(1-\alpha)^2 \delta_i^2}{M_f^2} \leq \| x_i - x_p^* \|^2 - \frac{(1-\alpha)^2 \delta_p^2}{M_f^2}. \end{aligned}$$

Summing up these inequalities in $i = k \dots p$ we get

$$(p+1-k)\frac{(1-\alpha)^2\delta_p^2}{M_f^2} \leq \| x_k - x_p^* \|^2 \leq D^2.$$

□

Note that the value $p+1-k$ is equal to the number of indices in the segment $[k,p]$. Now we can prove the efficiency estimate of the level method.

THEOREM 3.3.1 *Let* $\operatorname{diam} Q = D$. *Then the scheme of the level method terminates no sooner than after*

$$N = \left\lfloor \frac{M_f^2 D^2}{\epsilon^2 \alpha(1-\alpha)^2(2-\alpha)} \right\rfloor + 1$$

iterations. Termination criterion of the method guarantees $f_k^* - f^* \leq \epsilon$.

Proof: Assume that $\delta_k \geq \epsilon$, $0 \leq k \leq N$. Let us divide the indices on the groups in the *decreasing* order

$$\{N,\dots,0\} = I(0) \bigcup I(2) \bigcup \dots \bigcup I(m),$$

such that

$$I(j) = [p(j), k(j)], \quad p(j) \geq k(j), \quad j = 0 \dots m,$$

$$p(0) = N, \quad p(j+1) = k(j) + 1, \quad k(m) = 0,$$

$$\delta_{k(j)} \leq \tfrac{1}{1-\alpha}\delta_{p(j)} < \delta_{k(j)+1} \equiv \delta_{p(j+1)}.$$

Clearly, for $j \geq 0$ we have

$$\delta_{p(j+1)} \geq \frac{\delta_{p(j)}}{1-\alpha} \geq \frac{\delta_{p(0)}}{(1-\alpha)^{j+1}} \geq \frac{\epsilon}{(1-\alpha)^{j+1}}.$$

In view of Lemma 3.3.2, $n(j) = p(j) + 1 - k(j)$ is bounded:

$$n(j) \leq \frac{M_f^2 D^2}{(1-\alpha)^2 \delta_{p(j)}^2} \leq \frac{M_f^2 D^2}{\epsilon^2 (1-\alpha)^2}(1-\alpha)^{2j}.$$

Therefore

$$N = \sum_{j=0}^{m} n(j) \leq \frac{M_f^2 D^2}{\epsilon^2(1-\alpha)^2} \sum_{j=0}^{m} (1-\alpha)^{2j} \leq \frac{M_f^2 D^2}{\epsilon^2(1-\alpha)^2(1-(1-\alpha)^2)}.$$

□

Let us discuss the above efficiency estimate. Note that we can obtain the optimal value of the level parameter α from the following maximization problem:

$$(1-\alpha)^2(1-(1-\alpha)^2) \quad \to \quad \max_{\alpha \in [0,1]}.$$

Its solution is $\alpha^* = \frac{1}{2+\sqrt{2}}$. Under this choice we have the following efficiency estimate of the level method: $N \le \frac{4}{\epsilon^2} M_f^2 D^2$. Comparing this result with Theorem 3.2.1 we see that the level method is optimal *uniformly* in the dimension of the space of variables. Note that the analytical complexity bound of this method in *finite* dimension is not known.

One of the advantages of this method is that the gap $\delta_k = f_k^* - \hat{f}_k^*$ provides us with an *exact* estimate of current accuracy. Usually, this gap converges to zero much faster than in the worst case situation. For the majority of real life optimization problems the accuracy $\epsilon = 10^{-4} - 10^{-5}$ is obtained by the method after $3n - 4n$ iterations.

3.3.4 Constrained minimization

Let us demonstrate how we can use the models for solving constrained minimization problems. Consider the problem

$$\begin{aligned} &\min \quad f(x), \\ \text{s.t.} \quad &f_j(x) \le 0, \ j = 1 \dots m, \\ &x \in Q, \end{aligned} \tag{3.3.4}$$

where Q is a bounded closed convex set, and functions $f(x)$, $f_j(x)$ are Lipschitz continuous on Q.

Let us rewrite this problem as a problem with a single functional constraint. Denote $\bar{f}(x) = \max\limits_{1 \le j \le m} f_j(x)$. Then we obtain the equivalent problem

$$\begin{aligned} &\min \quad f(x), \\ \text{s.t.} \quad &\bar{f}(x) \le 0, \\ &x \in Q. \end{aligned} \tag{3.3.5}$$

Note that $f(x)$ and $\bar{f}(x)$ are convex and Lipschitz continuous. In this section we will try to solve (3.3.5) using the models for both of them.

Let us define the corresponding models. Consider a sequence $X = \{x_k\}_{k=0}^{\infty}$. Denote

$$\begin{aligned} \hat{f}_k(X; x) &= \max_{0 \le j \le k} [f(x_j) + \langle g(x_j), x - x_j \rangle] \le f(x), \\ \check{f}_k(X; x) &= \max_{0 \le j \le k} [\bar{f}(x_j) + \langle \bar{g}(x_j), x - x_j \rangle] \le \bar{f}(x), \end{aligned}$$

where $g(x_j) \in \partial f(x_j)$ and $\bar{g}(x_j) \in \partial \bar{f}(x_j)$.

As in Section 2.3.4, our scheme is based on the *parametric* function

$$\begin{aligned} f(t;x) &= \max\{f(x)-t, \bar{f}(x)\}, \\ f^*(t) &= \min_{x\in Q} f(t;x). \end{aligned}$$

Recall that $f^*(t)$ is nonincreasing in t. Let x^* be a solution to (3.3.5). Denote $t^* = f(x^*)$. Then t^* is the smallest root of function $f^*(t)$.

Using the models for the objective function and the constraint, we can introduce a model for the parametric function. Denote

$$\begin{aligned} f_k(X;t,x) &= \max\{\hat{f}_k(X;x)-t, \check{f}_k(X;x)\} \leq f(t;x), \\ \hat{f}_k^*(X;t) &= \min_{x\in Q} f_k(X;t,x) \leq f^*(t). \end{aligned}$$

Again, $\hat{f}_k^*(X;t)$ is nonincreasing in t. It is clear that its smallest root $t_k^*(X)$ does not exceed t^*.

We will need the following characterization of the root $t_k^*(X)$.

LEMMA 3.3.4

$$t_k^*(X) = \min\{\hat{f}_k(X;x) \mid \check{f}_k(X;x) \leq 0, \ x \in Q\}.$$

Proof: Denote by $\hat{x}_k^*$ the solution of the minimization problem in the right-hand side of the above equation. And let $\hat{t}_k^* = \hat{f}_k(X;\hat{x}_k^*)$. Then

$$\hat{f}_k^*(X;\hat{t}_k^*) \leq \max\{\hat{f}_k(X;\hat{x}_k^*) - \hat{t}_k^*, \check{f}_k(X;\hat{x}_k^*)\} \leq 0.$$

Thus, we always have $\hat{t}_k^* \geq t_k^*(X)$.

Assume that $\hat{t}_k^* > t_k^*(X)$. Then there exists a point y such that

$$\hat{f}_k(X;y) - t_k^*(X) \leq 0, \quad \check{f}_k(X;y) \leq 0.$$

However, in this case $\hat{t}_k^* = \hat{f}_k(X;\hat{x}_k^*) \leq \hat{f}_k(X;y) \leq t_k^*(X) < \hat{t}_k^*$. That is a contradiction. □

In our analysis we will need also the function

$$f_k^*(X;t) = \min_{0\leq j\leq k} f_k(X;t,x_j),$$

the *record value* of our parametric model.

LEMMA 3.3.5 *Let $t_0 < t_1 \le t^*$. Assume that $\hat{f}_k^*(X;t_1) > 0$. Then $t_k^*(X) > t_1$ and*

$$\hat{f}_k^*(X;t_0) \ge \hat{f}_k^*(X;t_1) + \tfrac{t_1 - t_0}{t_k^*(X) - t_1} \hat{f}_k^*(X;t_1). \qquad (3.3.6)$$

Proof. Denote $x_k^*(t) \in \operatorname{Arg\,min} f_k(X;t,x)$, $t_2 = t_k^*(X)$, $\alpha = \frac{t_1 - t_0}{t_2 - t_0} \in [0,1]$. Then

$$t_1 = (1-\alpha)t_0 + \alpha t_2$$

and inequality (3.3.6) is equivalent to the following:

$$\hat{f}_k^*(X;t_1) \le (1-\alpha)\hat{f}_k^*(X;t_0) + \alpha \hat{f}_k^*(X;t_2) \qquad (3.3.7)$$

(note that $\hat{f}_k^*(X;t_2) = 0$). Let $x_\alpha = (1-\alpha)x_k^*(t_0) + \alpha x_k^*(t_2)$. Then we have

$$\hat{f}_k^*(X;t_1) \le \max\{\hat{f}_k(X;x_\alpha) - t_1; \check{f}_k(X;x_\alpha)\}$$

$$\le \max\{(1-\alpha)(\hat{f}_k(X;x_k^*(t_0)) - t_0) + \alpha(\hat{f}_k(X;x_k^*(t_2)) - t_2);$$

$$(1-\alpha)\check{f}_k(X;x_k^*(t_0)) + \alpha \check{f}_k(X;x_k^*(t_2))\}$$

$$\le (1-\alpha)\max\{\hat{f}_k(X;x_k^*(t_0)) - t_0; \check{f}_k(X;x_k^*(t_0))\}$$

$$+\alpha \max\{\hat{f}_k(X;x_k^*(t_2)) - t_2; \check{f}_k(X;x_k^*(t_2))\}$$

$$= (1-\alpha)\hat{f}_k^*(X;t_0) + \alpha \hat{f}_k^*(X;t_2),$$

and we get (3.3.7). □

We also need the following statement (compare with Lemma 2.3.5).

LEMMA 3.3.6 *For any $\Delta \ge 0$ we have*

$$f^*(t) - \Delta \le f^*(t+\Delta),$$

$$\hat{f}_k^*(X;t) - \Delta \le \hat{f}_k^*(X;t+\Delta)$$

Proof. Indeed, for $f^*(t)$ we have

$$f^*(t+\Delta) = \min_{x\in Q} [\max\{f(x) - t; \bar{f}(x) + \Delta\} - \Delta]$$

$$\ge \min_{x\in Q} [\max\{f(x) - t; \bar{f}(x)\} - \Delta] = f^*(t) - \Delta.$$

The proof of the second inequality is similar. □

Now we are ready to present a constrained minimization scheme (compare with constrained minimization scheme of Section 2.3.5).

Constrained level method

0. Choose $x_0 \in Q$, $t_0 < t^*$, $\kappa \in (0, \frac{1}{2})$ and accuracy $\epsilon > 0$.

1. kth iteration ($k \geq 0$).

a). Keep generating sequence $X = \{x_j\}_{j=0}^{\infty}$ by the level method as applied to function $f(t_k; x)$. If the internal termination criterion (3.3.8)

$$\hat{f}_j^*(X; t_k) \geq (1 - \kappa) f_j^*(X; t_k)$$

holds, then stop the internal process and set $j(k) = j$.

Global stop: $f_j^*(X; t_k) \leq \epsilon$.

b). Set $t_{k+1} = t_{j(k)}^*(X)$.

We are interested in an analytical complexity bound for this method. Therefore the complexity of computation of the root $t_j^*(X)$ and of the value $\hat{f}_j^*(X; t)$ is not important for us now. We need to estimate the rate of convergence of the *master process* and the complexity of Step 1a).

Let us start from the master process.

LEMMA 3.3.7 *For all $k \geq 0$, we have*

$$f_{j(k)}^*(X; t_k) \leq \frac{t_0 - t^*}{1-\kappa} \left[\frac{1}{2(1-\kappa)}\right]^k.$$

Proof: Denote

$$\sigma_k = \frac{f_{j(k)}^*(X; t_k)}{\sqrt{t_{k+1} - t_k}}, \quad \beta = \frac{1}{2(1-\kappa)} \quad (< 1).$$

Since $t_{k+1} = t_{j(k)}^*(X)$ and in view of Lemma 3.3.5, for all $k \geq 1$, we have

$$\begin{aligned} \sigma_{k-1} &= \frac{1}{\sqrt{t_k - t_{k-1}}} f_{j(k-1)}^*(X; t_{k-1}) \geq \frac{1}{\sqrt{t_k - t_{k-1}}} \hat{f}_{j(k)}^*(X; t_{k-1}) \\ &\geq \frac{2}{\sqrt{t_{k+1} - t_k}} \hat{f}_{j(k)}^*(X; t_k) \geq \frac{2(1-\kappa)}{\sqrt{t_{k+1} - t_k}} f_{j(k)}^*(X; t_k) = \frac{\sigma_k}{\beta}. \end{aligned}$$

Thus, $\sigma_k \leq \beta\sigma_{k-1}$ and we obtain

$$\begin{aligned} f^*_{j(k)}(X;t_k) &= \sigma_k\sqrt{t_{k+1}-t_k} \leq \beta^k\sigma_0\sqrt{t_{k+1}-t_k} \\ &= \beta^k f^*_{j(0)}(X;t_0)\sqrt{\tfrac{t_{k+1}-t_k}{t_1-t_0}}. \end{aligned}$$

Further, in view of Lemma 3.3.6, $t_1 - t_0 \geq \hat{f}^*_{j(0)}(X;t_0)$. Therefore

$$\begin{aligned} f^*_{j(k)}(X;t_k) &\leq \beta^k f^*_{j(0)}(X;t_0)\sqrt{\tfrac{t_{k+1}-t_k}{\hat{f}^*_{j(0)}(X;t_0)}} \\ &\leq \tfrac{\beta^k}{1-\kappa}\sqrt{\hat{f}^*_{j(0)}(X;t_0)(t_{k+1}-t_k)} \leq \tfrac{\beta^k}{1-\kappa}\sqrt{f^*(t_0)(t_0-t^*)}. \end{aligned}$$

It remains to note that $f^*(t_0) \leq t_0 - t^*$ (see Lemma 3.3.6). □

Let Global stop condition in (3.3.8) be satisfied: $f^*_j(X;t_k) \leq \epsilon$. Then there exist j^* such that

$$f(t_k;x_{j^*}) = f^*_j(X;t_k) \leq \epsilon.$$

Therefore we have

$$f(t_k;x_{j^*}) = \max\{f(x_{j^*}) - t_k; \bar{f}(x_{j^*})\} \leq \epsilon.$$

Since $t_k \leq t^*$, we conclude that

$$\begin{aligned} f(x_{j^*}) &\leq t^* + \epsilon, \\ \bar{f}(x_{j^*}) &\leq \epsilon. \end{aligned} \tag{3.3.9}$$

In view of Lemma 3.3.7, we can get (3.3.9) at most in

$$N(\epsilon) = \tfrac{1}{\ln[2(1-\kappa)]}\ln\tfrac{t_0-t^*}{(1-\kappa)\epsilon}$$

full iterations of the master process. (The last iteration of the process is terminated by the Global stop rule). Note that in the above expression κ is an absolute constant (for example, we can take $\kappa = \frac{1}{4}$).

Let us estimate the complexity of the internal process. Denote

$$M_f = \max\{\| g \| \mid g \in \partial f(x) \bigcup \partial\bar{f}(x),\ x \in Q\}.$$

We need to analyze two cases.

1. *Full step.* At this step the internal process is terminated by the rule

$$\hat{f}^*_{j(k)}(X;t_k) \geq (1-\kappa)f^*_{j(k)}(X;t_k).$$

The corresponding inequality for the *gap* is as follows:

$$f^*_{j(k)}(X;t_k) - \hat{f}^*_{j(k)}(X;t_k) \leq \kappa f^*_{j(k)}(X;t_k).$$

In view of Theorem 3.3.1, this happens at most after

$$\frac{M_f^2 D^2}{\kappa^2 (f^*_{j(k)}(X;t_k))^2 \alpha(1-\alpha)^2(2-\alpha)}$$

iterations of the internal process. Since at the full step $f^*_{j(k)}(X;t_k)) \geq \epsilon$, we conclude that

$$j(k) - j(k-1) \leq \frac{M_f^2 D^2}{\kappa^2 \epsilon^2 \alpha(1-\alpha)^2(2-\alpha)}$$

for any full iteration of the master process.

2. *Last step.* The internal process of this step was terminated by Global stop rule:

$$f^*_j(X;t_k) \leq \epsilon.$$

Since the normal stopping criterion did not work, we conclude that

$$f^*_{j-1}(X;t_k) - \hat{f}^*_{j-1}(X;t_k) \geq \kappa f^*_{j-1}(X;t_k) \geq \kappa\epsilon.$$

Therefore, in view of Theorem 3.3.1, the number of iterations at the last step does not exceed

$$\frac{M_f^2 D^2}{\kappa^2 \epsilon^2 \alpha(1-\alpha)^2(2-\alpha)}.$$

Thus, we come to the following estimate of *total* complexity of the constrained level method:

$$(N(\epsilon)+1)\frac{M_f^2 D^2}{\kappa^2 \epsilon^2 \alpha(1-\alpha)^2(2-\alpha)}$$

$$= \frac{M_f^2 D^2}{\kappa^2 \epsilon^2 \alpha(1-\alpha)^2(2-\alpha)}\left[1 + \frac{1}{\ln[2(1-\kappa)]}\ln\frac{t_0 - t^*}{(1-\kappa)\epsilon}\right]$$

$$= \frac{M_f^2 D^2 \ln\frac{2(t_0-t^*)}{\epsilon}}{\epsilon^2 \alpha(1-\alpha)^2(2-\alpha)\kappa^2 \ln[2(1-\kappa)]}.$$

It can be shown that the reasonable choice for the parameters of this scheme is $\alpha = \kappa = \frac{1}{2+\sqrt{2}}$.

The principal term in the above complexity estimate is on the order of $\frac{1}{\epsilon^2}\ln\frac{2(t_0-t^*)}{\epsilon}$. Thus, the constrained level method is suboptimal (see Theorem 3.2.1).

In this method, at each iteration of the master process we need to find the root $t^*_{j(k)}(X)$. In view of Lemma 3.3.4, that is equivalent to the following problem:

$$\min\{\hat{f}_k(X;x) \mid \check{f}_k(X;x) \leq 0,\ x \in Q\}.$$

In other words, we need to solve the problem

$$\begin{aligned} &\min \quad t, \\ \text{s.t.} \quad & f(x_j) + \langle g(x_j), x - x_j \rangle \leq t, \quad j = 0 \dots k, \\ & \bar{f}(x_j) + \langle \bar{g}(x_j), x - x_j \rangle \leq 0, \quad j = 0 \dots k, \\ & x \in Q. \end{aligned}$$

If Q is a polytope, this problem can be solved by finite linear programming methods (simplex method). If Q is more complicated, we need to use interior-point schemes.

To conclude this section, let us note that we can use a better model for the functional constraints. Since

$$\bar{f}(x) = \max_{1 \leq i \leq m} f_i(x),$$

it is possible to work with

$$\check{f}_k(X; x) = \max_{0 \leq j \leq k} \max_{1 \leq i \leq m} [f_i(x_j) + \langle g_i(x_j), x - x_j \rangle],$$

where $g_i(x_j) \in \partial f_i(x_j)$. In practice, this *complete* model significantly accelerates the convergence of the process. However, clearly each iteration becomes more expensive.

As far as practical behavior of this scheme is concerned, we note that usually the process is very fast. There are some technical problems, related to accumulation of too many linear pieces in the model. However, in all practical schemes there exists some strategy for dropping the old elements of the model.

Chapter 4

STRUCTURAL OPTIMIZATION

4.1 Self-concordant functions

(Do we really have a black box? What the Newton method actually does? Definition of self-concordant functions; Main properties; Minimizing the self-concordant function.)

4.1.1 Black box concept in convex optimization

In this chapter we are going to present the main ideas underlying the modern polynomial-time interior-point methods in nonlinear optimization. In order to start, let us look first at the traditional formulation of a minimization problem.

Suppose we want to solve a minimization problem in the following form:

$$\min_{x \in R^n} \{ f_0(x) \mid f_j(x) \leq 0, \ j = 1 \dots m \}.$$

We assume that the functional components of this problem are convex. Note that all standard convex optimization schemes for solving this problem are based on the black-box concept. This means that we assume our problem to be equipped with an oracle, which provides us with some information on the functional components of the problem at some test point x. This oracle is local: If we change the shape of a component far enough from the test point, the answer of the oracle does not change. These answers comprise the only information available for numerical methods.[1]

However, if we look carefully at the above situation, we can see a certain contradiction. Indeed, in order to apply the convex optimization

[1] We have discussed this concept and the corresponding methods in the previous chapters.

methods, we need to be *sure* that our functional components are convex. However, we can check convexity only by analyzing the *structure* of these functions[2]: If our function is obtained from the *basic* convex functions by *convex* operations (summation, maximum, etc.), we conclude that it is convex.

Thus, the functional components of the problem are not in a black box at the moment we check their convexity and choose a minimization scheme. But we put them in a black box for numerical methods. That is the main conceptual contradiction of the standard convex optimization theory.[3]

The above observation gives us hope that the structure of the problem can be used to improve the performance of convex minimization schemes. Unfortunately, structure is a very fuzzy notion, which is quite difficult to formalize. One possible way to describe the structure is to fix the *analytical type* of functional components. For example, we can consider the problems with linear functions $f_j(x)$ only. This works, but note that this approach is very fragile: If we add just a single functional component of different type, we get another problem class and all theory must be done from scratch.

Alternatively, it is clear that having the structure at hand we can play a lot with the *analytical form* of the problem. We can rewrite the problem in many equivalent forms using nontrivial transformation of variables or constraints, introducing additional variables, etc. However, this would serve no purpose until the moment we realize the final goal of such transformations. So, let us try to find the goal.

At this moment, it is better to look at classical examples. In many situations the sequential reformulations of the initial problem can be seen as a part of the numerical scheme. We start from a complicated problem $\mathcal{P}$ and, step by step, we simplify its structure up to the moment we get a trivial problem (or, a problem which we know how to solve):

$$\mathcal{P} \longrightarrow \ldots \longrightarrow (f^*, x^*).$$

Let us look at the standard approach for solving a system of linear equations, namely,

$$Ax = b.$$

We can proceed as follows:

1. Check that A is symmetric and positive definite. Sometimes this is clear from the origin of matrix A.

[2] A numerical verification of convexity is a hopeless problem.

[3] However, the conclusions of the theory concerning the oracle-based minimization *schemes* remain valid.

2. Compute the Cholesky factorization of the matrix:

$$A = LL^T,$$

where L is a lower-triangular matrix. Form an auxiliary system

$$Ly = b, \quad L^T x = y.$$

3. Solve the auxiliary system.

This process can be seen as a sequence of equivalent transformations of the initial problem

Imagine for a moment that we do not know how to solve systems of linear equations. In order to discover the above scheme we should perform the following steps:

1. Find a class of problems which can be solved very efficiently (linear systems with triangular matrices in our example).

2. Describe the transformation rules for converting our initial problem into the desired form.

3. Describe the class of problems for which these transformation rules are applicable.

We are ready to explain the way it works in optimization. First of all, we need to find a *basic* numerical scheme and a problem formulation at which this scheme is very efficient. We will see that for our goals the most appropriate candidate is the *Newton method* (see Section 1.2.4) as applied in the framework of *Sequential Unconstrained Minimization* (see Section 1.3.3).

In the succeeding section we will highlight some drawbacks of the standard analysis of Newton method. From this analysis we derive a family of very special convex functions, the *self-concordant functions* and *self-concordant barriers*, which can be efficiently minimized by the Newton method. We use these objects in a description of a transformed version of our initial problem. In the sequel we refer to this description as to a *barrier model* of our problem. This model will replace the standard functional model of optimization problem used in the previous chapters.

4.1.2 What the Newton method actually does?

Let us look at the standard result on local convergence of the Newton method (we have proved it as Theorem 1.2.5). We are trying to find an unconstrained local minimum x^* of twice differentiable function $f(x)$. Assume that:

- $f''(x^*) \succeq lI_n$ with some constant $l > 0$,
- $\| f''(x) - f''(y) \| \leq M \| x - y \|$ for all x and $y \in R^n$.

We assume also that the starting point of the Newton process x_0 is close enough to x^*:

$$\| x_0 - x^* \| < \bar{r} = \tfrac{2l}{3M}. \qquad (4.1.1)$$

Then we can prove that the sequence

$$x_{k+1} = x_k - [f''(x_k]^{-1} f'(x_k), \quad k \geq 0, \qquad (4.1.2)$$

is well defined. Moreover, $\| x_k - x^* \| < \bar{r}$ for all $k \geq 0$ and the Newton method (4.1.2) converges quadratically:

$$\| x_{k+1} - x^* \| \leq \tfrac{M \| x_k - x^* \|^2}{2(l - M \| x_k - x^* \|)}.$$

What is wrong with this result? Note that the description of the *region* of quadratic convergence (4.1.1) for this method is given in terms of the *standard* inner product

$$\langle x, y \rangle = \sum_{i=1}^{n} x^{(i)} y^{(i)}.$$

If we choose a new basis in R^n, then all objects in our description change: the metric, the Hessians, the bounds l and M. But let us look what happens with the Newton process. Namely, let A be a nondegenerate $(n \times n)$-matrix. Consider the function

$$\phi(y) = f(Ay).$$

The following result is very important for understanding the nature of Newton method.

LEMMA 4.1.1 *Let $\{x_k\}$ be a sequence, generated by the Newton method for function f:*

$$x_{k+1} = x_k - [f''(x_k)]^{-1} f'(x_k), \quad k \geq 0.$$

Consider a sequence $\{y_k\}$, generated by the Newton method for function ϕ:

$$y_{k+1} = y_k - [\phi''(y_k)]^{-1} \phi'(y_k), \quad k \geq 0,$$

with $y_0 = A^{-1} x_0$. Then $y_k = A^{-1} x_k$ for all $k \geq 0$.

Proof: Let $y_k = A^{-1} x_k$ for some $k \geq 0$. Then

$$\begin{aligned} y_{k+1} &= y_k - [\phi''(y_k)]^{-1} \phi'(y_k) = y_k - [A^T f''(Ay_k) A]^{-1} A^T f'(Ay_k) \\ &= A^{-1} x_k - A^{-1} [f''(x_k)]^{-1} f'(x_k) = A^{-1} x_{k+1}. \end{aligned}$$

□

Thus, the Newton method is *affine invariant* with respect to affine transformation of variables. Therefore its real region of quadratic convergence *does not depend* on a particular inner product. It depends only on the local topological structure of function $f(x)$.

Let us try to understand what was bad in our assumptions. The main assumption we used is the Lipschitz continuity of Hessians:

$$\| f''(x) - f''(y) \| \leq M \| x - y \|, \quad \forall x, y \in R^n.$$

Let us assume that $f \in C^3(R^n)$. Denote

$$f'''(x)[u] = \lim_{\alpha \to 0} \tfrac{1}{\alpha}[f''(x + \alpha u) - f''(x)].$$

Note that the object in the right-hand side is an $(n \times n)$-matrix. Then our assumption is equivalent to

$$\| f'''(x)[u] \| \leq M \| u \| .$$

This means that at any point $x \in R^n$ we have

$$\langle f'''(x)[u]v, v\rangle \leq M \| u \| \cdot \| v \|^2 \quad \forall u, v \in R^n.$$

Note that the value in the left-hand side of this inequality is invariant with respect to affine transformation of variables. However, the right-hand side does not possess this property. Therefore the most natural way to improve the situation is to find an affine-invariant replacement for the standard norm $\| \cdot \|$. The main candidate for such a replacement is rather evident: That is the norm defined by the Hessian $f''(x)$ itself, namely,

$$\| u \|_{f''(x)} = \langle f''(x)u, u\rangle^{1/2}.$$

This choice gives us the class of *self-concordant functions*.

4.1.3 Definition of self-concordant function

Let us consider a *closed convex* function $f(x) \in C^3(\mathrm{dom}\, f)$ with *open* domain. Let us fix a point $x \in \mathrm{dom}\, f$ and a direction $u \in R^n$. Consider the function

$$\phi(x; t) = f(x + tu),$$

as a function of variable $t \in \mathrm{dom}\, \phi(x; \cdot) \subseteq R^1$. Denote

$$\begin{aligned} Df(x)[u] &= \phi'(x; t) = \langle f'(x), u\rangle, \\ D^2 f(x)[u, u] &= \phi''(x; t) = \langle f''(x)u, u\rangle = \| u \|^2_{f''(x)}, \\ D^3 f(x)[u, u, u] &= \phi'''(x; t) = \langle D^3 f(x)[u]u, u\rangle. \end{aligned}$$

DEFINITION 4.1.1 *We call function f* self-concordant *if there exists a constant $M_f \geq 0$ such that the inequality*

$$D^3 f(x)[u, u, u] \leq M_f \parallel u \parallel_{f''(x)}^{3/2}$$

holds for any $x \in \operatorname{dom} f$ and $u \in R^n$.

Note that we cannot expect these functions to be very widespread. But we need them only to construct a barrier model of our problem. We will see very soon that such functions are easy to be minimized by the Newton method.

Let us point out an equivalent definition of self-concordant functions.

LEMMA 4.1.2 *A function f is self-concordant if and only if for any $x \in \operatorname{dom} f$ and any u_1, u_2, $u_3 \in R^n$ we have*

$$\mid D^3 f(x)[u_1, u_2, u_3] \mid \leq M_f \prod_{i=1}^{3} \parallel u_i \parallel_{f''(x)} . \qquad (4.1.3)$$

We accept this statement without proof since it needs some special facts from the theory of three-linear symmetric forms.

In what follows, we very often use Definition 4.1.1 in order to prove that some f is self-concordant. On the contrary, Lemma 4.1.2 is useful for establishing the properties of self-concordant functions.

Let us consider several examples.

EXAMPLE 4.1.1 1. *Linear function.* Consider the function

$$f(x) = \alpha + \langle a, x\rangle, \quad \operatorname{dom} f = R^n.$$

Then

$$f'(x) = a, \quad f''(x) = 0, \quad f'''(x) = 0,$$

and we conclude that $M_f = 0$.

2. *Convex quadratic function.* Consider the function

$$f(x) = \alpha + \langle a, x\rangle + \tfrac{1}{2}\langle Ax, x\rangle, \quad \operatorname{dom} f = R^n,$$

where $A = A^T \succeq 0$. Then

$$f'(x) = a + Ax, \quad f''(x) = A, \quad f'''(x) = 0,$$

and we conclude that $M_f = 0$.

3. *Logarithmic barrier for a ray.* Consider a function of one variable

$$f(x) = -\ln x, \quad \operatorname{dom} f = \{x \in R^1 \mid x > 0\}.$$

Then

$$f'(x) = -\tfrac{1}{x}, \quad f''(x) = \tfrac{1}{x^2}, \quad f'''(x) = -\tfrac{2}{x^3}.$$

Therefore $f(x)$ is self-concordant with $M_f = 2$.

4. *Logarithmic barrier for a second-order region.* Let $A = A^T \succeq 0$. Consider the *concave* function

$$\phi(x) = \alpha + \langle a, x\rangle - \tfrac{1}{2}\langle Ax, x\rangle.$$

Define $f(x) = -\ln \phi(x)$, with $\operatorname{dom} f = \{x \in R^n \mid \phi(x) > 0\}$. Then

$$\begin{aligned}
Df(x)[u] &= -\tfrac{1}{\phi(x)}[\langle a, u\rangle - \langle Ax, u\rangle],\\
D^2 f(x)[u,u] &= \tfrac{1}{\phi^2(x)}[\langle a, u\rangle - \langle Ax, u\rangle]^2 + \tfrac{1}{\phi(x)}\langle Au, u\rangle,\\
D^3 f(x)[u,u,u] &= -\tfrac{2}{\phi^3(x)}[\langle a, u\rangle - \langle Ax, u\rangle]^3\\
&\quad -\tfrac{3}{\phi^2(x)}[\langle a, u\rangle - \langle Ax, u\rangle]\langle Au, u\rangle.
\end{aligned}$$

Denote $\omega_1 = Df(x)[u]$ and $\omega_2 = \frac{1}{\phi(x)}\langle Au, u\rangle$. Then

$$\begin{aligned}
D^2 f(x)[u,u] &= \omega_1^2 + \omega_2 \geq 0,\\
\mid D^3 f(x)[u,u,u] \mid &= \mid 2\omega_1^3 + 3\omega_1\omega_2 \mid .
\end{aligned}$$

The only nontrivial case is $\omega_1 \neq 0$. Denote $\alpha = \omega_2/\omega_1^2$. Then

$$\tfrac{|D^3 f(x)[u,u,u]|}{(D^2 f(x)[u,u])^{3/2}} \leq \tfrac{2|\omega_1|^3 + 3|\omega_1|\omega_2}{(\omega_1^2+\omega_2)^{3/2}} = \tfrac{2(1+\frac{3}{2}\alpha)}{(1+\alpha)^{3/2}} \leq 2.$$

Thus, this function is self-concordant and $M_f = 2$.

5. It is easy to verify that none of the following functions of one variable is self-concordant:

$$f(x) = e^x, \quad f(x) = \tfrac{1}{x^p},\ x > 0,\ p > 0, \quad f(x) = \mid x \mid^p,\ p > 2.$$

□

Let us look now at the main properties of self-concordant functions.

THEOREM 4.1.1 *Let functions f_i be self-concordant with constants M_i, $i = 1, 2$, and let α, $\beta > 0$. Then the function $f(x) = \alpha f_1(x) + \beta f_2(x)$ is self-concordant with constant*

$$M_f = \max\left\{\frac{1}{\sqrt{\alpha}}M_1, \frac{1}{\sqrt{\beta}}M_2\right\}$$

and $\operatorname{dom} f = \operatorname{dom} f_1 \bigcap \operatorname{dom} f_2$.

Proof: In view of Theorem 3.1.5, f is a closed convex function. Let us fix some $x \in \operatorname{dom} f$ and $u \in R^n$. Then

$$| D^3 f_i(x)[u, u, u] | \leq M_i \left[D^2 f_i(x)[u, u]\right]^{3/2}, \quad i = 1, 2.$$

Denote $\omega_i = D^2 f_i(x)[u, u] \geq 0$. Then

$$\begin{aligned} \frac{|D^3 f(x)[u,u,u]|}{[D^2 f(x)[u,u]]^{3/2}} &\leq \frac{\alpha |D^3 f_1(x)[u,u,u]| + \beta |D^3 f_2(x)[u,u,u]|}{[\alpha D^1 f_1(x)[u,u] + \beta D^2 f_2(x)[u,u]]^{3/2}} \\ &\leq \frac{\alpha M_1 \omega_1^{3/2} + \beta M_2 \omega_2^{3/2}}{[\alpha\omega_1 + \beta\omega_2]^{3/2}}. \end{aligned}$$

The right-hand side of this inequality does not change when we replace (ω_1, ω_2) by $(t\omega_1, t\omega_2)$ with $t > 0$. Therefore we can assume that

$$\alpha\omega_1 + \beta\omega_2 = 1.$$

Denote $\xi = \alpha\omega_1$. Then the right-hand side of the above inequality becomes equal to

$$\frac{M_1}{\sqrt{\alpha}}\xi^{3/2} + \frac{M_2}{\sqrt{\beta}}(1 - \xi)^{3/2}, \quad \xi \in [0, 1].$$

This function is convex in ξ. Therefore it attains its maximum at the end points of the interval (see Corollary 3.1.1). □

COROLLARY 4.1.1 *Let function f be self-concordant with some constant M_f. If $A = A^T \succeq 0$, then the function*

$$\phi(x) = \alpha + \langle a, x\rangle + \tfrac{1}{2}\langle Ax, x\rangle + f(x)$$

is also self-concordant with constant $M_\phi = M_f$.

Proof: We have seen that any convex quadratic function is self-concordant with the constant equal to zero. □

COROLLARY 4.1.2 *Let function f be self-concordant with some constant M_f and $\alpha > 0$. Then the function $\phi(x) = \alpha f(x)$ is also self-concordant with the constant $M_\phi = \frac{1}{\sqrt{\alpha}} M_f$.* □

Let us prove now that self-concordance is an affine-invariant property.

THEOREM 4.1.2 *Let $\mathcal{A}(x) = Ax + b$: $R^n \to R^m$, be a linear operator. Assume that function $f(y)$ is self-concordant with constant M_f. Then the function $\phi(x) = f(\mathcal{A}(x))$ is also self-concordant and $M_\phi = M_f$.*

Proof: The function $\phi(x)$ is closed and convex in view of Theorem 3.1.6. Let us fix some $x \in \operatorname{dom} \phi = \{x : \mathcal{A}(x) \in \operatorname{dom} f\}$ and $u \in R^n$. Denote $y = \mathcal{A}(x)$, $v = Au$. Then

$$\begin{aligned} D\phi(x)[u] &= \langle f'(\mathcal{A}(x)), Au\rangle = \langle f'(y), v\rangle, \\ D^2\phi(x)[u,u] &= \langle f''(\mathcal{A}(x))Au, Au\rangle = \langle f''(y)v, v\rangle, \\ D^3\phi(x)[u,u,u] &= D^3 f(\mathcal{A}(x))[Au, Au, Au] = D^3 f(y)[v,v,v]. \end{aligned}$$

Therefore

$$\begin{aligned} \mid D^3\phi(x)[u,u,u] \mid &= \mid D^3 f(y)[v,v,v] \mid \leq M_f \langle f''(y)v, v\rangle^{3/2} \\ &= M_f (D^2\phi(x)[u,u])^{3/2}. \end{aligned}$$

□

The next statement demonstrates that some local properties of a self-concordant function reflect somehow the global properties of its domain.

THEOREM 4.1.3 *Let function f be self-concordant. If* $\operatorname{dom} f$ *contains no straight line, then the Hessian $f''(x)$ is nondegenerate at any x from* $\operatorname{dom} f$.

Proof: Assume that $\langle f''(x)u, u\rangle = 0$ for some $x \in \operatorname{dom} f$ and $u \in R^n$, $u \neq 0$. Consider the points $y_\alpha = x + \alpha u \in \operatorname{dom} f$ and the function

$$\psi(\alpha) = \langle f''(y_\alpha)u, u\rangle.$$

Note that

$$\psi'(\alpha) = D^3 f(y_\alpha)[u,u,u] \leq 2\psi(\alpha)^{3/2}, \quad \psi(0) = 0.$$

Since $\psi(\alpha) \geq 0$, we conclude that $\psi'(0) = 0$. Therefore this function is a part of the solution of the following system of differential equations:

$$\psi(0) = \xi(0) = 0, \quad \left\{ \begin{aligned} \psi'(\alpha) &= 2\psi(\alpha)^{3/2} - \xi(\alpha), \\ \xi'(\alpha) &= 0. \end{aligned} \right.$$

However, this system has a unique trivial solution. Therefore $\psi(\alpha) = 0$ for all feasible α.

Thus, we have shown that the function $\phi(\alpha) = f(y_\alpha)$ is linear:

$$\begin{aligned} \phi(\alpha) &= f(x) + \langle f'(x), y_\alpha - x\rangle + \int\limits_0^\alpha \int\limits_0^\lambda \langle f''(y_\tau)u, u\rangle d\tau d\lambda \\ &= f(x) + \alpha \langle f'(x), u\rangle. \end{aligned}$$

Assume that there exists $\bar{\alpha}$ such that $y_{\bar{\alpha}} \in \partial(\mathrm{dom}\, f)$. Consider a sequence $\{\alpha_k\}$ such that $\alpha_k \uparrow \bar{\alpha}$. Then

$$z_k = (y_{\alpha_k}, \phi(\alpha_k)) \to \bar{z} = (y_{\bar{\alpha}}, \phi(\bar{\alpha})).$$

Note that $z_k \in \mathrm{epi}\, f$, but $\bar{z} \notin \mathrm{epi}\, f$ since $y_{\bar{\alpha}} \notin \mathrm{dom}\, f$. That is a contradiction since function f is closed. Considering direction $-u$, and assuming that this ray intersects the boundary, we come to a contradiction again. Therefore we conclude that $y_\alpha \in \mathrm{dom}\, f$ for all α. However, that is a contradiction with the assumptions of the theorem. □

Finally, let us describe the behavior of self-concordant function near the boundary of its domain.

THEOREM 4.1.4 *Let f be a self-concordant function. Then for any point $\bar{x} \in \partial(\mathrm{dom}\, f)$ and any sequence*

$$\{x_k\} \subset \mathrm{dom}\, f: \quad x_k \to \bar{x}$$

we have $f(x_k) \to +\infty$.

Proof: Note that the sequence $\{f(x_k)\}$ is bounded below:

$$f(x_k) \geq f(x_0) + \langle f'(x_0), x_k - x_0\rangle.$$

Assume that it is bounded from above. Then it has a limit point $\bar{f}$. Of course, we can think that this is a unique limit point of the sequence. Therefore

$$z_k = (x_k, f(x_k)) \to \bar{z} = (\bar{x}, \bar{f}).$$

Note that $z_k \in \mathrm{epi}\, f$, but $\bar{z} \notin \mathrm{epi}\, f$ since $\bar{x} \notin \mathrm{dom}\, f$. That is a contradiction since function f is closed. □

Thus, we have proved that $f(x)$ is a *barrier function* for $\mathrm{cl}\,(\mathrm{dom}\, f)$ (see Section 1.3.3).

4.1.4 Main inequalities

Let us fix some self-concordant function $f(x)$. We assume that its constant $M_f = 2$ (otherwise we can scale it, see Corollary 4.1.2). We call such functions the *standard* self-concordant. We assume also that $\operatorname{dom} f$ contains no straight line (this implies that all $f''(x)$ are nondegenerate, see Theorem 4.1.3).

Denote:

$$\begin{aligned} \| u \|_x &= \langle f''(x)u, u\rangle^{1/2}, \\ \| v \|_x^* &= \langle [f''(x)]^{-1}v, v\rangle^{1/2}, \\ \lambda_f(x) &= \langle [f''(x)]^{-1}f'(x), f'(x)\rangle^{1/2}. \end{aligned}$$

Clearly, $| \langle v, u\rangle | \leq \| v \|_x^* \cdot \| u \|_x$. We call $\| u \|_x$ the *local norm* of direction u with respect to x, and $\lambda_f(x) = \| f'(x) \|_x^*$ is called the *local norm of the gradient* $f'(x)$.[4]

Let us fix $x \in \operatorname{dom} f$ and $u \in R^n$, $u \neq 0$. Consider the function of one variable

$$\phi(t) = \frac{1}{\langle f''(x+tu)u,u\rangle^{1/2}}$$

with the domain $\operatorname{dom}\phi = \{t \in R^1 : \ x + tu \in \operatorname{dom} f\}$.

LEMMA 4.1.3 *For all feasible t we have $| \phi'(t) | \leq 1$.*

Proof: Indeed,

$$\phi'(t) = -\frac{f'''(x+tu)[u,u,u]}{2\langle f''(x+tu)u,u\rangle^{3/2}}.$$

Therefore $| \phi'(t) | \leq 1$ in view of Definition 4.1.1. □

COROLLARY 4.1.3 *Domain of function $\phi(t)$ contains the interval*

$$(-\phi(0), \phi(0)).$$

Proof: Since $f(x+tu) \to \infty$ as $x+tu$ approaches the boundary of $\operatorname{dom} f$ (see Theorem 4.1.4), the function $\langle f''(x + tu)u, u\rangle$ cannot be bounded. Therefore $\operatorname{dom}\phi \equiv \{t \mid \ \phi(t) > 0\}$. It remains to note that

$$\phi(t) \geq \phi(0) - | t |$$

in view of Lemma 4.1.3. □

---

[4]Sometimes $\lambda_f(x)$ is called the *Newton decrement* of function f at x.

Let us consider the following ellipsoid:

$$W^0(x;r) = \{y \in R^n \mid \| y - x \|_x < r\},$$

$$W(x;r) = \text{cl}\,(W^0(x;r)) \equiv \{y \in R^n \mid \| y - x \|_x \leq r\}.$$

This ellipsoid is called the *Dikin* ellipsoid of function f at x.

THEOREM 4.1.5 *1. For any $x \in \operatorname{dom} f$ we have $W^0(x;1) \subseteq \operatorname{dom} f$.*
2. For all x, $y \in \operatorname{dom} f$ the following inequality holds:

$$\| y - x \|_y \geq \frac{\|y-x\|_x}{1+\|y-x\|_x}. \qquad (4.1.4)$$

3. If $\| y - x \|_x < 1$, then

$$\| y - x \|_y \leq \frac{\|y-x\|_x}{1-\|y-x\|_x}. \qquad (4.1.5)$$

Proof: 1. In view of Corollary 4.1.3, $\operatorname{dom} f$ contains the set

$$\{y = x + tu \mid t^2 \| u \|_x^2 < 1\}$$

(since $\phi(0) = 1/\| u \|_x$). That is exactly $W^0(x;1)$.

2. Let us choose $u = y - x$. Then

$$\phi(1) = \frac{1}{\|y-x\|_y}, \quad \phi(0) = \frac{1}{\|y-x\|_x},$$

and $\phi(1) \leq \phi(0) + 1$ in view of Lemma 4.1.3. That is (4.1.4).

3. If $\| y - x \|_x < 1$, then $\phi(0) > 1$, and in view of Lemma 4.1.3 $\phi(1) \geq \phi(0) - 1$. That is (4.1.5). □

THEOREM 4.1.6 *Let $x \in \operatorname{dom} f$. Then for any $y \in W^0(x;1)$ we have*

$$(1 - \| y - x \|_x)^2 f''(x) \preceq f''(y) \preceq \frac{1}{(1-\|y-x\|_x)^2} f''(x). \qquad (4.1.6)$$

Proof: Let us fix some $u \in R^n$, $u \neq 0$. Consider the function

$$\psi(t) = \langle f''(x + t(y-x))u, u \rangle, \quad t \in [0,1].$$

Denote $y_t = x + t(y - x)$. Then, in view of Lemma 4.1.2 and (4.1.5), we have

$$\begin{aligned} \mid \psi'(t) \mid &= \mid D^3 f(y_t)[y-x, u, u] \mid \leq 2 \| y - x \|_{y_t} \| u \|_{y_t}^2 \\ &= \frac{2}{t} \| y_t - x \|_{y_t} \psi(t) \leq \frac{2}{t} \cdot \frac{\|y_t - x\|_x}{1-\|y_t-x\|_x} \cdot \psi(t) \\ &= \frac{2\|y-x\|_x}{1-t\|y-x\|_x} \cdot \psi(t). \end{aligned}$$

Therefore

$$2(\ln(1 - t \parallel y - x \parallel_x))' \leq (\ln \psi(t))' \leq -2(\ln(1 - t \parallel y - x \parallel_x))'.$$

Let us integrate this inequality in $t \in [0, 1]$. We get:

$$(1 - \parallel y - x \parallel_x)^2 \leq \frac{\psi(1)}{\psi(0)} \leq \frac{1}{(1 - \parallel y - x \parallel_x)^2}.$$

That is exactly (4.1.6). □

COROLLARY 4.1.4 *Let $x \in \operatorname{dom} f$ and $r = \parallel y - x \parallel_x < 1$. Then we can estimate the matrix*

$$G = \int_0^1 f''(x + \tau(y - x))d\tau$$

as follows:

$$(1 - r + \tfrac{r^2}{3})f''(x) \preceq G \preceq \tfrac{1}{1-r} f''(x).$$

Proof: Indeed, in view of Theorem 4.1.6 we have

$$\begin{aligned} G &= \int_0^1 f''(x + \tau(y - x))d\tau \succeq f''(x) \cdot \int_0^1 (1 - \tau r)^2 d\tau \\ &= (1 - r + \tfrac{1}{3} r^2) f''(x), \\ G &\preceq f''(x) \cdot \int_0^1 \frac{d\tau}{(1 - \tau r)^2} = \tfrac{1}{1-r} f''(x). \end{aligned}$$

□

Let us look again at the most important facts we have proved.

- At any point $x \in \operatorname{dom} f$ we can point out an *ellipsoid*

  $$W^0(x; 1) = \{x \in R^n \mid \langle f''(x)(y - x), y - x \rangle < 1\},$$

  belonging to $\operatorname{dom} f$.

- Inside the ellipsoid $W(x; r)$ with $r \in [0, 1)$ function f is almost quadratic:

  $$(1 - r)^2 f''(x) \preceq f''(y) \preceq \tfrac{1}{(1-r)^2} f''(x)$$

  for all $y \in W(x; r)$. Choosing r small enough, we can make the quality of the quadratic approximation acceptable for our goals.

These two facts form the basis for almost all consequent results.

We conclude this section with the results describing the variation of a self-concordant function with respect to a linear approximation.

THEOREM 4.1.7 *For any x, $y \in \operatorname{dom} f$ we have*

$$\langle f'(y) - f'(x), y - x\rangle \geq \frac{\|y-x\|_x^2}{1+\|y-x\|_x}, \qquad (4.1.7)$$

$$f(y) \geq f(x) + \langle f'(x), y - x\rangle + \omega(\| y - x \|_x), \qquad (4.1.8)$$

where $\omega(t) = t - \ln(1+t)$.

Proof: Denote $y_\tau = x + \tau(y-x)$, $\tau \in [0,1]$, and $r = \| y - x \|_x$. Then, in view of (4.1.4) we have

$$\begin{aligned}
\langle f'(y) - f'(x), y - x\rangle &= \int_0^1 \langle f''(y_\tau)(y-x), y-x\rangle d\tau \\
&= \int_0^1 \frac{1}{\tau^2} \| y_\tau - x \|_{y_\tau}^2 \, d\tau \\
&\geq \int_0^1 \frac{r^2}{(1+\tau r)^2} d\tau = r \int_0^r \frac{1}{(1+t)^2} dt = \frac{r^2}{1+r}.
\end{aligned}$$

Further, using (4.1.7), we obtain

$$\begin{aligned}
f(y) - f(x) - \langle f'(x), y - x\rangle &= \int_0^1 \langle f'(y_\tau) - f'(x), y - x\rangle d\tau \\
&= \int_0^1 \frac{1}{\tau}\langle f'(y_\tau) - f'(x), y_\tau - x\rangle d\tau \\
&\geq \int_0^1 \frac{\|y_\tau - x\|_x^2}{\tau(1+\|y_\tau - x\|_x)} d\tau = \int_0^1 \frac{\tau r^2}{1+\tau r} d\tau \\
&= \int_0^r \frac{t dt}{1+t} = \omega(r).
\end{aligned}$$

□

THEOREM 4.1.8 *Let $x \in \operatorname{dom} f$ and $\| y - x \|_x < 1$. Then*

$$\langle f'(y) - f'(x), y - x\rangle \leq \frac{\|y-x\|_x^2}{1-\|y-x\|_x}, \qquad (4.1.9)$$

$$f(y) \leq f(x) + \langle f'(x), y - x\rangle + \omega_*(\| y - x \|_x), \qquad (4.1.10)$$

where $\omega_*(t) = -t - \ln(1-t)$.

Proof: Denote $y_\tau = x + \tau(y-x)$, $\tau \in [0,1]$, and $r = \parallel y - x \parallel_x$. Since $\parallel y_\tau - x \parallel < 1$, in view of (4.1.5) we have

$$\begin{aligned}
\langle f'(y) - f'(x), y - x\rangle &= \int_0^1 \langle f''(y_\tau)(y-x), y-x\rangle d\tau \\
&= \int_0^1 \tfrac{1}{\tau^2} \parallel y_\tau - x \parallel_{y_\tau}^2 d\tau \\
&\leq \int_0^1 \tfrac{r^2}{(1-\tau r)^2} d\tau = r \int_0^r \tfrac{1}{(1-t)^2} dt = \tfrac{r^2}{1-r}.
\end{aligned}$$

Further, using (4.1.9), we obtain

$$\begin{aligned}
f(y) - f(x) - \langle f'(x), y-x\rangle &= \int_0^1 \langle f'(y_\tau) - f'(x), y - x\rangle d\tau \\
&= \int_0^1 \tfrac{1}{\tau}\langle f'(y_\tau) - f'(x), y_\tau - x\rangle d\tau \\
&\leq \int_0^1 \tfrac{\|y_\tau - x\|_x^2}{\tau(1-\|y_\tau - x\|_x)} d\tau = \int_0^1 \tfrac{\tau r^2}{1-\tau r} d\tau \\
&= \int_0^r \tfrac{t dt}{1-t} = \omega_*(r).
\end{aligned}$$

□

THEOREM 4.1.9 *Inequalities (4.1.4), (4.1.5), (4.1.7), (4.1.8), (4.1.9) and (4.1.10) are necessary and sufficient characteristics of standard self-concordant functions.*

Proof: We have justified two sequences of implications:

$$\text{Definition 4.1.1} \quad \Rightarrow \quad (4.1.4) \quad \Rightarrow \quad (4.1.7) \quad \Rightarrow \quad (4.1.8),$$

$$\text{Definition 4.1.1} \quad \Rightarrow \quad (4.1.5) \quad \Rightarrow \quad (4.1.9) \quad \Rightarrow \quad (4.1.10).$$

Let us prove the implication (4.1.8) $\Rightarrow$ Definition 4.1.1. Let $x \in \operatorname{dom} f$ and $x - \alpha u \in \operatorname{dom} f$ for $\alpha \in [0, \epsilon)$. Consider the function

$$\psi(\alpha) = f(x - \alpha u), \quad \alpha \in [0, \epsilon).$$

Denote $r = \|u\|_x \equiv [\phi''(0)]^{1/2}$. Assuming that (4.1.8) holds for all x and y from $\mathrm{dom}\, f$, we have

$$\psi(\alpha) - \psi(0) - \psi'(0)\alpha - \tfrac{1}{2}\psi''(0)\alpha^2 \geq \omega(\alpha r) - \tfrac{1}{2}\alpha^2 r^2.$$

Therefore

$$\begin{aligned}
\tfrac{1}{6}\psi'''(0) &= \lim_{\alpha\downarrow 0}\left[\psi(\alpha) - \psi(0) - \psi'(0)\alpha - \tfrac{1}{2}\psi''(0)\alpha^2\right] \\
&\geq \lim_{\alpha\downarrow 0}\tfrac{1}{\alpha^3}\left[\omega(\alpha r) - \tfrac{1}{2}\alpha^2 r^2\right] = \lim_{\alpha\downarrow 0}\tfrac{r}{3\alpha^2}\left[\omega'(\alpha r) - \alpha r\right] \\
&= \lim_{\alpha\downarrow 0}\tfrac{r}{3\alpha^2}\left[\tfrac{\alpha r}{1+\alpha r} - \alpha r\right] = -\tfrac{r^3}{3}.
\end{aligned}$$

Thus, $D^3 f(x)[u,u,u] = -\psi'''(0) \leq \psi'''(0) \leq 2[\psi''(0)]^{3/2}$ and that is Definition 4.1.1 with $M_f = 2$. Implication (4.1.10) $\Rightarrow$ Definition 4.1.1 can be proved by a similar reasoning. □

The above theorems are written in terms of two auxiliary functions $\omega(t) = t - \ln(1+t)$ and $\omega_*(\tau) = -\tau - \ln(1-\tau)$. Note that

$$\begin{aligned}
\omega'(t) &= \tfrac{t}{1+t} \geq 0, & \omega''(t) &= \tfrac{1}{(1+t)^2} > 0, \\
\omega_*'(\tau) &= \tfrac{\tau}{1-\tau} \geq 0, & \omega_*''(\tau) &= \tfrac{1}{(1-\tau)^2} > 0.
\end{aligned}$$

Therefore, $\omega(t)$ and $\omega_*(\tau)$ are convex functions. In what follows we often use different relations between these functions. Let us fix this notation for future references.

LEMMA 4.1.4 *For any $t \geq 0$ and $\tau \in [0,1)$ we have*

$$\omega'(\omega_*'(\tau)) = \tau, \quad \omega_*'(\omega'(t)) = t,$$

$$\omega(t) = \max_{0\leq\xi<1}[\xi t - \omega_*(\xi)], \quad \omega_*(\tau) = \max_{\xi\geq 0}[\xi\tau - \omega(\xi)],$$

$$\omega(t) + \omega_*(\tau) \geq \tau t,$$

$$\omega_*(\tau) = \tau\omega_*'(\tau) - \omega(\omega_*'(\tau)), \quad \omega(t) = t\omega'(t) - \omega_*(\omega'(t)).$$

We leave the proof of this lemma as an exercise for the reader. For an advanced reader we should note that the only reason for the above relations is that functions $\omega(t)$ and $\omega_*(t)$ are *conjugate.*

Let us prove two more inequalities.

THEOREM 4.1.10 *For any x and y from Q we have*

$$f(y) \geq f(x) + \langle f'(x), y - x\rangle + \omega(\| f'(y) - f'(x)\|_y^*). \tag{4.1.11}$$

If in addition $\| f'(y) - f'(x)\|_y^* < 1$, then

$$f(y) \leq f(x) + \langle f'(x), y - x\rangle + \omega_*(\| f'(y) - f'(x)\|_y^*). \tag{4.1.12}$$

Proof: Let us fix an arbitrary x and y from Q. Consider the function

$$\phi(z) = f(z) - \langle f'(x), z\rangle, \quad z \in Q.$$

Note that this function is self-concordant and $\phi'(x) = 0$. Therefore, using inequality (4.1.10) we get

$$\begin{aligned} f(x) - \langle f'(x), x\rangle &= \phi(x) = \min_{z\in Q} \phi(z) \\ &\leq \min_{z\in Q}[\phi(y) + \langle \phi'(y), z - y\rangle + \omega_*(\|z - y\|_y)] \\ &= \phi(y) - \omega(\|\phi'(y)\|_y^*) \\ &= f(y) - \langle f'(x), y\rangle - \omega(\| f'(y) - f'(x)\|_y^*), \end{aligned}$$

and that is (4.1.11). In order to prove inequality (4.1.12) we use a similar reasoning with (4.1.8). □

4.1.5 Minimizing the self-concordant function

Let us consider the following minimization problem:

$$\min\{f(x) \mid x \in \operatorname{dom} f\}. \tag{4.1.13}$$

The next theorem provides us with a sufficient condition for existence of its solution. Recall that we assume that f is a standard self-concordant function and $\operatorname{dom} f$ contains no straight line.

THEOREM 4.1.11 *Let $\lambda_f(x) < 1$ for some $x \in \operatorname{dom} f$. Then the solution of problem (4.1.13), x_f^*, exists and is unique.*

Proof: Indeed, in view of (4.1.8), for any $y \in \operatorname{dom} f$ we have

$$\begin{aligned} f(y) &\geq f(x) + \langle f'(x), y - x\rangle + \omega(\| y - x \|_x) \\ &\geq f(x) - \| f'(x) \|_x^* \cdot \| y - x \|_x + \omega(\| y - x \|_x) \\ &= f(x) - \lambda_f(x) \cdot \| y - x \|_x + \omega(\| y - x \|_x). \end{aligned}$$

Therefore for any $y \in \mathcal{L}_f(f(x)) = \{y \in R^n \mid f(y) \leq f(x)\}$ we have

$$\frac{1}{\| y-x \|_x}\omega(\| y - x \|_x) \leq \lambda_f(x) < 1.$$

Note that the function $\frac{1}{t}\omega(t) = 1 - \frac{1}{t}\ln(1+t)$ is strictly increasing in t. Hence, $\| y - x \|_x \leq \bar{t}$, where $\bar{t}$ is a unique positive root of the equation

$$(1 - \lambda_f(x))t = \ln(1+t).$$

Thus, $\mathcal{L}_f(f(x))$ is bounded and therefore x_f^* exists. It is unique since in view of (4.1.8) for all $y \in \mathrm{dom}\, f$ we have

$$f(y) \geq f(x_f^*) + \omega(\| y - x_f^* \|_{x_f^*}).$$

□

Thus, we have proved that a local condition $\lambda_f(x) < 1$ provides us with some global information on function f, that is the existence of the minimum x_f^*. Note that the result of Theorem 4.1.11 cannot be strengthened.

EXAMPLE 4.1.2 Let us fix some $\epsilon > 0$. Consider a function of one variable

$$f_\epsilon(x) = \epsilon x - \ln x, \quad x > 0.$$

This function is self-concordant in view of Example 4.1.1 and Corollary 4.1.1. Note that

$$f_\epsilon'(x) = \epsilon - \frac{1}{x}, \quad f_\epsilon'' = \frac{1}{x^2}.$$

Therefore $\lambda_{f_\epsilon}(x) = \mid 1 - \epsilon x \mid$. Thus, for $\epsilon = 0$ we have $\lambda_{f_0}(x) = 1$ for any $x > 0$. Note that the function f_0 is not bounded below.

If $\epsilon > 0$, then $x_{f_\epsilon}^* = \frac{1}{\epsilon}$. Note that we can recognize the existence of the minimizer at point $x = 1$ even if ϵ is arbitrary small. □

Let us consider now a scheme of the *damped Newton method*:

Damped Newton method
0. Choose $x_0 \in \mathrm{dom}\, f$.   **1.** Iterate $x_{k+1} = x_k - \frac{1}{1+\lambda_f(x_k)}[f''(x_k)]^{-1}f'(x_k)$, $k \geq 0$.

(4.1.14)

THEOREM 4.1.12 *For any $k \geq 0$ we have*

$$f(x_{k+1}) \leq f(x_k) - \omega(\lambda_f(x_k)). \qquad (4.1.15)$$

Proof: Denote $\lambda = \lambda_f(x_k)$. Then $\| x_{k+1} - x_k \|_x = \frac{\lambda}{1+\lambda} = \omega'(\lambda)$. Therefore, in view of (4.1.10) and Lemma 4.1.4, we have

$$\begin{aligned} f(x_{k+1}) &\leq f(x_k) + \langle f'(x_k), x_{k+1} - x_k \rangle + \omega_*(\| x_{k+1} - x_k \|_x) \\ &= f(x_k) - \frac{\lambda^2}{1+\lambda} + \omega_*(\omega'(\lambda)) \\ &= f(x_k) - \lambda\omega'(\lambda) + \omega_*(\omega'(\lambda)) = f(x_k) - \omega(\lambda). \end{aligned}$$

□

Thus, for all $x \in \operatorname{dom} f$ with $\lambda_f(x) \geq \beta > 0$ one step of the damped Newton method decreases the value of $f(x)$ at least by a constant $\omega(\beta) > 0$. Note that the result of Theorem 4.1.12 is *global*. It can be used to obtain a global efficiency estimate of the process.

Let us describe now the *local* convergence of the *standard* Newton method:

Standard Newton method
0. Choose $x_0 \in \operatorname{dom} f$. **1.** Iterate $x_{k+1} = x_k - [f''(x_k)]^{-1} f'(x_k)$, $k \geq 0$.

(4.1.16)

Note that we can measure the convergence of this process in different ways. We can estimate the rate of convergence for the *functional gap* $f(x_k) - f(x_f^*)$, or for the local norm of the gradient $\lambda_f(x_k) = \| f'(x_k) \|_{x_k}^*$, or for the *local distance to the minimum* $\| x_k - x_f^* \|_{x_k}$. Finally, we can look at the distance to the minimum in a fixed metrics

$$r_*(x_k) \equiv \|x_k - x_f^*\|_{x_f^*}$$

defined by the minimum itself. Let us prove that locally all these measures are equivalent.

THEOREM 4.1.13 *Let $\lambda_f(x) < 1$. Then*

$$\omega(\lambda_f(x)) \leq f(x) - f(x_f^*) \leq \omega_*(\lambda_f(x)), \tag{4.1.17}$$

$$\omega'(\lambda_f(x)) \leq \parallel x - x_f^* \parallel_x \leq \omega_*'(\lambda_f(x)), \tag{4.1.18}$$

$$\omega(r_*(x)) \leq f(x) - f(x_f^*) \leq \omega_*(r_*(x)), \tag{4.1.19}$$

where the last inequality is valid for $r_*(x) < 1$.

Proof: Denote $r = \parallel x - x_f^* \parallel_x$ and $\lambda = \lambda_f(x)$. Inequalities (4.1.17) follow from Theorem 4.1.10. Further, in view of (4.1.7) we have

$$\tfrac{r^2}{1+r} \leq \langle f'(x), x - x_f^* \rangle \leq \lambda r.$$

That is the right-hand side of inequality (4.1.18). If $r \geq 1$ then the left-hand side of this inequality is trivial. Suppose that $r < 1$. Then $f'(x) = G(x - x_f^*)$ with

$$G = \int_0^1 f''(x_f^* + \tau(x - x_f^*))d\tau,$$

and

$$\lambda_f^2(x) = \langle [f''(x)]^{-1} G(x - x_f^*), G(x - x_f^*) \rangle \leq \parallel H \parallel^2 r^2,$$

where $H = [f''(x)]^{-1/2} G [f''(x)]^{-1/2}$. In view of Corollary 4.1.4, we have

$$G \preceq \tfrac{1}{1-r} f''(x).$$

Therefore $\parallel H \parallel \leq \frac{1}{1-r}$ and we conclude that

$$\lambda_f^2(x) \leq \tfrac{r^2}{(1-r)^2} = (\omega_*'(r))^2.$$

Thus, $\lambda_f(x) \leq \omega_*'(r)$. Applying $\omega'(\cdot)$ to both sides, we get the remaining part of (4.1.18).

Finally, inequalities (4.1.19) follow from (4.1.8) and (4.1.10). □

Let us estimate the local rate convergence of the standard Newton method (4.1.16). It is convenient to do that in terms of $\lambda_f(x)$, the local norm of the gradient.

THEOREM 4.1.14 *Let $x \in \operatorname{dom} f$ and $\lambda_f(x) < 1$. Then the point*

$$x_+ = x - [f''(x)]^{-1} f'(x)$$

belongs to dom f *and we have*

$$\lambda_f(x_+) \leq \left(\frac{\lambda_f(x)}{1-\lambda_f(x)}\right)^2.$$

Proof: Denote $p = x_+ - x$, $\lambda = \lambda_f(x)$. Then $\| p \|_x = \lambda < 1$. Therefore $x_+ \in \operatorname{dom} f$ (see Theorem 4.1.5). Note that in view of Theorem 4.1.6,

$$\begin{aligned} \lambda_f(x_+) &= \langle [f''(x_+)]^{-1} f'(x_+), f'(x_+)\rangle^{1/2} \\ &\leq \frac{1}{1-\|p\|_x} \| f'(x_+) \|_x = \frac{1}{1-\lambda} \| f'(x_+) \|_x . \end{aligned}$$

Further,

$$f'(x_+) = f'(x_+) - f'(x) - f''(x)(x_+ - x) = Gp,$$

where $G = \int\limits_0^1 [f''(x + \tau p) - f''(x)] d\tau$. Therefore

$$\| f'(x_+) \|_x^2 = \langle [f''(x)]^{-1} Gp, Gp\rangle \leq \| H \|^2 \cdot \| p \|_x^2,$$

where $H = [f''(x)]^{-1/2} G [f''(x)]^{-1/2}$. In view of Corollary 4.1.4,

$$(-\lambda + \tfrac{1}{3}\lambda^2) f''(x) \preceq G \preceq \tfrac{\lambda}{1-\lambda} f''(x).$$

Therefore $\| H \| \leq \max\left\{\frac{\lambda}{1-\lambda}, \lambda - \frac{1}{3}\lambda^2\right\} = \frac{\lambda}{1-\lambda}$, and we conclude that

$$\lambda_f^2(x_+) \leq \frac{1}{(1-\lambda)^2} \| f'(x_+) \|_x^2 \leq \frac{\lambda^4}{(1-\lambda)^4}.$$

□

Theorem 4.1.14 provides us with the following description of the region of quadratic convergence of scheme (4.1.16):

$$\lambda_f(x) < \bar{\lambda} = \frac{3-\sqrt{5}}{2} = 0.3819\ldots,$$

where $\bar{\lambda}$ is the root of the equation $\frac{\lambda}{(1-\lambda)^2} = 1$. In this case we can guarantee that $\lambda_f(x_+) < \lambda_f(x)$.

Thus, our results lead to the following strategy for solving the initial problem (4.1.13).

- *First stage*: $\lambda_f(x_k) \geq \beta$, where $\beta \in (0, \bar{\lambda})$. At this stage we apply the damped Newton method. At each iteration of this method we have

  $$f(x_{k+1}) \leq f(x_k) - \omega(\beta).$$

  Thus, the number of steps of this stage is bounded:

  $$N \leq \frac{1}{\omega(\beta)}[f(x_0) - f(x_f^*)].$$

- *Second stage:* $\lambda_f(x_k) \leq \beta$. At this stage we apply the standard Newton method. This process converges quadratically:

$$\lambda_f(x_{k+1}) \leq \left(\frac{\lambda_f(x_k)}{1-\lambda_f(x_k)}\right)^2 \leq \frac{\beta\lambda_f(x_k)}{(1-\beta)^2} < \lambda_f(x_k).$$

It can be shown that the local convergence of the damped Newton method (4.1.14) is also quadratic:

$$x_+ = x - \frac{[f''(x)]^{-1}f'(x)}{1+\lambda_f(x)} \quad \Rightarrow \quad \lambda_f(x_+) \leq 2\lambda_f^2(x). \tag{4.1.20}$$

However, we prefer to use the above switching strategy since it gives better complexity bounds. Relation (4.1.20) can be justified in the same way as it was done in Theorem 4.1.14. We leave the reasoning as an exercise for the reader.

4.2 Self-concordant barriers

(Motivation; Definition of self-concordant barriers; Main properties; Standard minimization problem; Central path; Path-following method; How to initialize the process? Problems with functional constraints.)

4.2.1 Motivation

In the previous section we have seen that the Newton method is very efficient in minimizing a standard *self-concordant* function. Such a function is always a barrier for its domain. Let us check what can be proved about the sequential unconstrained minimization approach (Section 1.3.3), which uses such barriers.

In what follows we deal with constrained minimization problems of special type. Denote $\mathrm{Dom}\, f = \mathrm{cl}\,(\mathrm{dom}\, f)$.

DEFINITION 4.2.1 *We call a constrained minimization problem* standard *if it has the form*

$$\min\{\langle c, x\rangle \mid x \in Q\}, \tag{4.2.1}$$

where Q is a closed convex set. We assume also that we know a self-concordant function f such that $\mathrm{Dom}\, f = Q$.

Let us introduce a parametric penalty function

$$f(t; x) = t\langle c, x\rangle + f(x)$$

with $t \geq 0$. Note that $f(t;x)$ is self-concordant in x (see Corollary 4.1.1). Denote

$$x^*(t) = \arg \min_{x\in \mathrm{dom}\, f} f(t;x).$$

This trajectory is called *the central path* of the problem (4.2.1). Note that we can expect $x^*(t) \to x^*$ as $t \to \infty$ (see Section 1.3.3). Therefore we are going to follow this trajectory.

Recall that the standard Newton method, as applied to minimization of function $f(t;x)$, has a local quadratic convergence (Theorem 4.1.14). Moreover, we have an explicit description of the region of quadratic convergence:

$$\lambda_{f(t;\cdot)}(x) \leq \beta < \bar{\lambda} = \tfrac{3-\sqrt{5}}{2}.$$

Let us study our possibilities assuming that we know exactly $x = x^*(t)$ for some $t > 0$.

Thus, we are going to increase t:

$$t_+ = t + \Delta, \quad \Delta > 0.$$

However, we need to keep x in the region of quadratic convergence of the Newton method for function $f(t + \Delta; \cdot)$:

$$\lambda_{f(t+\Delta;\cdot)}(x) \leq \beta < \bar{\lambda}.$$

Note that the update $t \to t_+$ does not change the Hessian of the barrier function:

$$f''(t + \Delta; x) = f''(t; x).$$

Therefore it is easy to estimate how can be big the step Δ. Indeed, the first order optimality condition provides us with the following *central path equation:*

$$tc + f'(x^*(t)) = 0. \tag{4.2.2}$$

Since $tc + f'(x) = 0$, we obtain

$$\lambda_{f(t+\Delta;\cdot)}(x) = \| t_+c + f'(x) \|_x = \Delta \| c \|_x = \tfrac{\Delta}{t} \| f'(x) \|_x \leq \beta.$$

Hence, if we want to increase t at a *linear rate*, we need to assume that the value

$$\lambda_f^2(x) = \| f'(x) \|_x^2 \equiv \langle [f''(x)]^{-1} f'(x), f'(x) \rangle$$

is *uniformly bounded* on $\operatorname{dom} f$.

Thus, we come to a definition of *self-concordant barrier.*

4.2.2 Definition of self-concordant barriers

DEFINITION 4.2.2 *Let $F(x)$ be a standard self-concordant function. We call it a ν-*self-concordant barrier *for set* Dom F, *if*

$$\sup_{u \in R^n} [2\langle F'(x), u\rangle - \langle F''(x)u, u\rangle] \leq \nu \tag{4.2.3}$$

for all $x \in \mathrm{dom}\, F$. *The value* ν *is called the* parameter *of the barrier.*

Note that we do not assume $F''(x)$ to be nondegenerate. However, if this is the case, then the inequality (4.2.3) is equivalent to

$$\langle [F''(x)]^{-1} F'(x), F'(x) \rangle \leq \nu. \tag{4.2.4}$$

We will use also another equivalent form of inequality (4.2.3):

$$\langle F'(x), u \rangle^2 \leq \nu \langle F''(x) u, u \rangle \quad \forall u \in R^n. \tag{4.2.5}$$

(To see that for u with $\langle F''(x)u, u\rangle > 0$, replace u in (4.2.3) by λu and find the maximum of the left-hand side in λ.) Note that the condition (4.2.5) can be written in a matrix notation:

$$F''(x) \succeq \tfrac{1}{\nu} F'(x) F'(x)^T. \tag{4.2.6}$$

Let us check now which self-concordant functions given by Example 4.1.1 are also self-concordant barriers.

EXAMPLE 4.2.1 1. *Linear function*: $f(x) = \alpha + \langle a, x\rangle$, $\mathrm{dom}\, f = R^n$. Clearly, for $a \neq 0$ this function is not a self-concordant barrier since $f''(x) = 0$.

2. *Convex quadratic function.* Let $A = A^T \succ 0$. Consider the function

$$f(x) = \alpha + \langle a, x\rangle + \tfrac{1}{2}\langle Ax, x\rangle, \quad \mathrm{dom}\, f = R^n.$$

Then $f'(x) = a + Ax$ and $f''(x) = A$. Therefore

$$\begin{aligned} \langle [f(x)]^{-1} f'(x), f'(x)\rangle &= \langle A^{-1}(Ax - a), Ax - a\rangle \\ &= \langle Ax, x\rangle - 2\langle a, x\rangle + \langle A^{-1}a, a\rangle. \end{aligned}$$

Clearly, this value is unbounded from above on R^n. Thus, a quadratic function is not a self-concordant barrier.

3. *Logarithmic barrier for a ray.* Consider the following function of one variable:

$$F(x) = -\ln x, \quad \mathrm{dom}\, F = \{x \in R^1 \mid x > 0\}.$$

Then $F'(x) = -\frac{1}{x}$ and $F''(x) = \frac{1}{x^2} > 0$. Therefore

$$\frac{(F'(x))^2}{F''(x)} = \frac{1}{x^2} \cdot x^2 = 1.$$

Thus, $F(x)$ is a ν-self-concordant barrier for $\{x > 0\}$ with $\nu = 1$.

4. *Logarithmic barrier for a second-order region.* Let $A = A^T \succeq 0$. Consider the *concave* quadratic function

$$\phi(x) = \alpha + \langle a, x\rangle - \tfrac{1}{2}\langle Ax, x\rangle.$$

Define $F(x) = -\ln \phi(x)$, $\operatorname{dom} f = \{x \in R^n \mid \phi(x) > 0\}$. Then

$$\begin{aligned}\langle F'(x), u\rangle &= -\tfrac{1}{\phi(x)}[\langle a, u\rangle - \langle Ax, u\rangle],\\ \langle F''(x)u, u\rangle &= \tfrac{1}{\phi^2(x)}[\langle a, u\rangle - \langle Ax, u\rangle]^2 + \tfrac{1}{\phi(x)}\langle Au, u\rangle.\end{aligned}$$

Denote $\omega_1 = \langle F'(x), u\rangle$ and $\omega_2 = \frac{1}{\phi(x)}\langle Au, u\rangle$. Then

$$\langle F''(x)u, u\rangle = \omega_1^2 + \omega_2 \geq \omega_1^2.$$

Therefore $2\langle F'(x), u\rangle - \langle F''(x)u, u\rangle \leq 2\omega_1 - \omega_1^2 \leq 1$. Thus, $F(x)$ is a ν-self-concordant barrier with $\nu = 1$. □

Let us present some simple properties of self-concordant barriers.

THEOREM 4.2.1 *Let $F(x)$ be a self-concordant barrier. Then the function $\langle c, x\rangle + F(x)$ is a self-concordant function on* $\operatorname{dom} F$.

Proof: Since $F(x)$ is a self-concordant function, we just apply Corollary 4.1.1. □

Note that this property is important for path-following schemes.

THEOREM 4.2.2 *Let F_i be ν_i-self-concordant barriers, $i = 1, 2$. Then the function*

$$F(x) = F_1(x) + F_2(x)$$

is a self-concordant barrier for convex set $\operatorname{Dom} F = \operatorname{Dom} F_1 \bigcap \operatorname{Dom} F_2$ *with the parameter $\nu = \nu_1 + \nu_2$.*

Proof: In view of Theorem 4.1.1, F is a standard self-concordant function. Let us fix $x \in \mathrm{dom}\, F$. Then

$$\begin{aligned}
& \max_{u\in R^n} [2\langle F'(x)u, u\rangle - \langle F''(x)u, u\rangle] \\
= & \max_{u\in R^n} [2\langle F_1'(x)u, u\rangle - \langle F_1''(x)u, u\rangle + 2\langle F_1'(x)u, u\rangle - \langle F_1''(x)u, u\rangle] \\
\leq & \max_{u\in R^n} [2\langle F_1'(x)u, u\rangle - \langle F_1''(x)u, u\rangle] \\
& + \max_{u\in R^n} [2\langle F_2'(x)u, u\rangle - \langle F_2''(x)u, u\rangle] \leq \nu_1 + \nu_2.
\end{aligned}$$

□

Finally, let us show that the value of a parameter of a self-concordant barrier is invariant with respect to affine transformation of variables.

THEOREM 4.2.3 *Let $\mathcal{A}(x) = Ax + b$ be a linear operator, $\mathcal{A}(x) : R^n \to R^m$. Assume that function $F(y)$ is a ν-self-concordant barrier. Then the function $\Phi(x) = F(\mathcal{A}(x))$ is a ν-self-concordant barrier for the set*

$$\mathrm{Dom}\, \Phi = \{x \in R^n \mid \mathcal{A}(x) \in \mathrm{Dom}\, F\}.$$

Proof: Function $\Phi(x)$ is a standard self-concordant function in view of Theorem 4.1.2. Let us fix $x \in \mathrm{dom}\, \Phi$. Then $y = \mathcal{A}(x) \in \mathrm{dom}\, F$. Note that for any $u \in R^n$ we have

$$\langle \Phi'(x), u\rangle = \langle F'(y), Au\rangle, \quad \langle \Phi''(x)u, u\rangle = \langle F''(y)Au, Au\rangle.$$

Therefore

$$\begin{aligned}
& \max_{u\in R^n} [2\langle \Phi'(x), u\rangle - \langle \Phi''(x)u, u\rangle] \\
& = \max_{u\in R^n} [2\langle F'(y), Au\rangle - \langle F''(y)Au, Au\rangle] \\
& \leq \max_{v\in R^m} [2\langle F'(y), v\rangle - \langle F''(y)v, v\rangle] \leq \nu.
\end{aligned}$$

□

4.2.3 Main inequalities

Let us show that the local characteristics of a self-concordant barrier (the gradient and the Hessian) provide us with *global* information about the structure of the domain.

THEOREM 4.2.4 *1. Let $F(x)$ be a ν-self-concordant barrier. For any x and y from* $\mathrm{dom}\, F$, *we have*

$$\langle F'(x), y - x\rangle < \nu. \qquad (4.2.7)$$

Moreover, if $\langle F'(x), y - x\rangle \geq 0$, *then*

$$\langle F'(y) - F'(x), y - x\rangle \geq \frac{\langle F'(x), y-x\rangle^2}{\nu - \langle F'(x), y-x\rangle}. \qquad (4.2.8)$$

2. A standard self-concordant function $F(x)$ *is a* ν*-self-concordant barrier if and only if*

$$F(y) \geq F(x) - \nu \ln\left(1 - \tfrac{1}{\nu}\langle F'(x), y - x\rangle\right) \quad \forall x, y \in \operatorname{dom} F. \qquad (4.2.9)$$

Proof: 1. Let $x, y \in \operatorname{dom} F$. Consider the function

$$\phi(t) = \langle F'(x + t(y - x)), y - x\rangle, \quad t \in [0, 1].$$

If $\phi(0) \leq 0$, then (4.2.7) is trivial. If $\phi(0) = 0$, then (4.2.8) is trivial. Suppose that $\phi(0) > 0$. Note that in view of (4.2.5) we have

$$\begin{aligned} \phi'(t) &= \langle F''(x + t(y - x))(y - x), y - x\rangle \\ &\geq \tfrac{1}{\nu}\langle F'(x + t(y - x)), y - x\rangle^2 = \tfrac{1}{\nu}\phi^2(t). \end{aligned}$$

Therefore $\phi(t)$ increases and it is positive for $t \in [0, 1]$. Moreover, for any $t \in [0, 1]$ we have

$$-\frac{1}{\phi(t)} + \frac{1}{\phi(0)} \geq \frac{1}{\nu}t.$$

This implies that $\langle F'(x), y - x\rangle = \phi(0) < \frac{\nu}{t}$ for all $t \in [0, 1]$. Thus, (4.2.7) is proved. Moreover,

$$\phi(t) - \phi(0) \geq \frac{\nu\phi(0)}{\nu - t\phi(0)} - \phi(0) = \frac{t\phi(0)^2}{\nu - t\phi(0)}, \quad t \in [0, 1].$$

Taking $t = 1$, we get (4.2.8).

2. Denote $\psi(x) = e^{-\frac{1}{\nu}F(x)}$. Then

$$\begin{aligned} \psi'(x) &= -\tfrac{1}{\nu}e^{-\frac{1}{\nu}F(x)} \cdot F'(x), \\ \psi''(x) &= -\tfrac{1}{\nu}e^{-\frac{1}{\nu}F(x)}\left[F''(x) - \tfrac{1}{\nu}F'(x)F'(x)^T\right]. \end{aligned}$$

Thus, in view of Theorem 2.1.4 and definition (4.2.6), function $\psi(x)$ is concave if and only if the function $F(x)$ is a ν-self-concordant barrier. It remains to note that (4.2.9) is the same as

$$\psi(y) \leq \psi(x) + \langle \psi'(x), y - x\rangle$$

up to a logarithmic transformation of both sides of the inequality. □

THEOREM 4.2.5 *Let $F(x)$ be a ν-self-concordant barrier. Then for any $x \in \operatorname{dom} F$ and $y \in \operatorname{Dom} F$ such that*

$$\langle F'(x), y - x \rangle \geq 0, \tag{4.2.10}$$

we have

$$\| y - x \|_x \leq \nu + 2\sqrt{\nu}. \tag{4.2.11}$$

Proof: Denote $r = \| y - x \|_x$. Let $r > \sqrt{\nu}$. Consider the point $y_\alpha = x + \alpha(y - x)$ with $\alpha = \frac{\sqrt{\nu}}{r} < 1$. In view of our assumption (4.2.10) and inequality (4.1.7) we have

$$\begin{aligned} \omega \equiv \langle F'(y_\alpha), y - x \rangle &\geq \langle F'(y_\alpha) - F'(x), y - x \rangle \\ &= \tfrac{1}{\alpha} \langle F'(y_\alpha) - F'(x), y_\alpha - x \rangle \\ &\geq \tfrac{1}{\alpha} \cdot \tfrac{\|y_\alpha - x\|_x^2}{1 + \|y_\alpha - x\|_x^2} = \tfrac{\alpha \|y - x\|_x^2}{1 + \alpha \|y - x\|_x} = \tfrac{r\sqrt{\nu}}{1 + \sqrt{\nu}}. \end{aligned}$$

On the other hand, in view of (4.2.7), we obtain

$$(1 - \alpha)\omega = \langle F'(y_\alpha), y - y_\alpha \rangle \leq \nu.$$

Thus,

$$\left(1 - \tfrac{\sqrt{\nu}}{r}\right) \tfrac{r\sqrt{\nu}}{1 + \sqrt{\nu}} \leq \nu,$$

and that is exactly (4.2.11). □

We conclude this section by studying the properties of one special point of a convex set.

DEFINITION 4.2.3 *Let $F(x)$ be a ν-self-concordant barrier for the set* $\operatorname{Dom} F$. *The point*

$$x_F^* = \arg \min_{x \in \operatorname{dom} F} F(x),$$

is called the analytic center *of convex set* $\operatorname{Dom} F$, *generated by the barrier* $F(x)$.

THEOREM 4.2.6 *Assume that the analytic center of a ν-self-concordant barrier $F(x)$ exists. Then for any $x \in \operatorname{Dom} F$ we have*

$$\| x - x_F^* \|_{x_F^*} \leq \nu + 2\sqrt{\nu}.$$

On the other hand, for any $x \in R^n$ such that $\| x - x_F^* \|_{x_F^*} \leq 1$ we have $x \in \mathrm{Dom}\, F$.

Proof: The first statement follows from Theorem 4.2.5 since $F'(x_F^*) = 0$. The second statement follows from Theorem 4.1.5. □

Thus, the *asphericity* of the set $\mathrm{Dom}\, F$ with respect to x_F^*, computed in the metric $\| \cdot \|_{x_F^*}$, does not exceed $\nu + 2\sqrt{\nu}$. It is well known that for any convex set in R^n there exists a metric in which the asphericity of this set is less than or equal to n (John Theorem). However, we managed to estimate the asphericity in terms of the *parameter* of the barrier. This value does not depend directly on the dimension of the space.

Note also, that if $\mathrm{Dom}\, F$ contains no straight line, the existence of x_F^* implies the boundedness of $\mathrm{Dom}\, F$. (Since then $F''(x_F^*)$ is nondegenerate, see Theorem 4.1.3).

COROLLARY 4.2.1 *Let* $\mathrm{Dom}\, F$ *be bounded. Then for any* $x \in \mathrm{dom}\, F$ *and* $v \in R^n$ *we have*

$$\| v \|_x^* \leq (\nu + 2\sqrt{\nu}) \| v \|_{x_F^*}^* .$$

Proof: By Lemma 3.1.12 we get the following representation:

$$\| v \|_x^* \equiv \langle [F''(x)]^{-1} v, v \rangle^{1/2} = \max\{\langle v, u \rangle \mid \langle F''(x) u, u \rangle \leq 1\}.$$

On the other hand, in view of Theorem 4.1.5 and Theorem 4.2.6, we have

$$\begin{aligned} B &\equiv \{y \in R^n \mid \| y - x \|_x \leq 1\} \subseteq \mathrm{Dom}\, F \\ &\subseteq \{y \in R^n \mid \| y - x_F^* \|_x \leq \nu + 2\sqrt{\nu}\} \equiv B_*. \end{aligned}$$

Therefore, using again Theorem 4.2.6, we get the following:

$$\begin{aligned} \| v \|_x^* &= \max\{\langle v, y - x \rangle \mid y \in B\} \leq \max\{\langle v, y - x \rangle \mid y \in B_*\} \\ &= \langle v, x_F^* - x \rangle + (\nu + 2\sqrt{\nu}) \| v \|_{x_F^*}^* . \end{aligned}$$

Note that $\| v \|_x^* = \| -v \|_x^*$. Therefore we can assume $\langle v, x_F^* - x \rangle \leq 0$. □

4.2.4 Path-following scheme

Now we are ready to describe a *barrier model* of the minimization problem. This is the *standard* minimization problem

$$\min\{\langle c, x \rangle \mid x \in Q\} \tag{4.2.12}$$

with *bounded* closed convex set $Q \equiv \text{Dom}\, F$, which has nonempty interior, and which is endowed with a ν-self-concordant barrier $F(x)$.

Recall that we are going to solve (4.2.12) by tracing the *central path*:

$$x^*(t) = \arg \min_{x \in \text{dom}\, F} f(t;x), \tag{4.2.13}$$

where $f(t;x) = t\langle c, x\rangle + F(x)$ and $t \geq 0$. In view of the first-order optimality condition, any point of the central path satisfies equation

$$tc + F'(x^*(t)) = 0. \tag{4.2.14}$$

Since the set Q is bounded, the *analytic center* of this set, x^*_F, exists and

$$x^*(0) = x^*_F. \tag{4.2.15}$$

In order to follow the central path, we are going to update the points, satisfying an *approximate centering condition*:

$$\lambda_{f(t;\cdot)}(x) \equiv \| f'(t;x) \|^*_x = \| tc + F'(x) \|^*_x \leq \beta, \tag{4.2.16}$$

where the *centering parameter* β is small enough.

Let us show that this is a reasonable goal.

THEOREM 4.2.7 *For any $t > 0$ we have*

$$\langle c, x^*(t)\rangle - c^* \leq \tfrac{\nu}{t}, \tag{4.2.17}$$

where c^* is the optimal value of (4.2.12). If a point x satisfies the centering condition (4.2.16), then

$$\langle c, x\rangle - c^* \leq \tfrac{1}{t}\left(\nu + \tfrac{(\beta+\sqrt{\nu})\beta}{1-\beta}\right). \tag{4.2.18}$$

Proof: Let x^* be a solution to (4.2.12). In view of (4.2.14) and (4.2.7) we have

$$\langle c, x^*(t) - x^*\rangle = \tfrac{1}{t}\langle F'(x^*(t)), x^* - x^*(t)\rangle \leq \tfrac{\nu}{t}.$$

Further, let x satisfy (4.2.16). Denote $\lambda = \lambda_{f(t;\cdot)}(x)$. Then

$$\begin{aligned} t\langle c, x - x^*(t)\rangle &= \langle f'(t;x) - F'(x), x - x^*(t)\rangle \\ &\leq (\lambda + \sqrt{\nu}) \| x - x^*(t) \|_x \\ &\leq (\lambda + \sqrt{\nu}) \tfrac{\lambda}{1-\lambda} \leq \tfrac{(\beta+\sqrt{\nu})\beta}{1-\beta} \end{aligned}$$

in view of (4.2.4), Theorem 4.1.13 and (4.2.16). □

Let us analyze now one step of a path-following scheme. Namely, assume that $x \in \operatorname{dom} F$. Consider the following iterate:

$$\boxed{\begin{aligned} t_+ &= t + \frac{\gamma}{\| c \|_x^*}, \\ x_+ &= x - [F''(x)]^{-1}(t_+ c + F'(x)). \end{aligned}} \qquad (4.2.19)$$

THEOREM 4.2.8 *Let x satisfy (4.2.16):*

$$\| tc + F'(x) \|_x^* \leq \beta$$

with $\beta < \bar{\lambda} = \frac{3-\sqrt{5}}{2}$. Then for γ, such that

$$| \gamma | \leq \frac{\sqrt{\beta}}{1+\sqrt{\beta}} - \beta, \qquad (4.2.20)$$

we have again $\| t_+ c + F'(x_+) \|_x^* \leq \beta$.

Proof: Denote $\lambda_0 = \| tc + F'(x) \|_x^* \leq \beta$, $\lambda_1 = \| t_+ c + F'(x) \|_x^*$ and $\lambda_+ = \| t_+ c + F'(x_+) \|_{x_+}^*$. Then

$$\lambda_1 \leq \lambda_0 + | \gamma | \leq \beta + | \gamma |$$

and in view of Theorem 4.1.14 we have

$$\lambda_+ \leq \left(\frac{\lambda_1}{1-\lambda_1} \right)^2 \equiv [\omega_*'(\lambda_1)]^2.$$

It remains to note that inequality (4.2.20) is equivalent to

$$\omega_*'(\beta + | \gamma |) \leq \sqrt{\beta}.$$

(recall that $\omega'(\omega_*'(\tau)) = \tau$, see Lemma 4.1.4). □

Let us prove now that the increase of t in the scheme (4.2.19) is sufficiently large.

LEMMA 4.2.1 *Let x satisfy (4.2.16). Then*

$$\| c \|_x^* \leq \frac{1}{t}(\beta + \sqrt{\nu}). \qquad (4.2.21)$$

Proof: Indeed, in view of (4.2.16) and (4.2.4), we have

$$t \parallel c \parallel_x^* = \parallel f'(t;x) - F'(x) \parallel_x^* \leq \parallel f'(t;x) \parallel_x^* + \parallel F'(x) \parallel_x^* \leq \beta + \sqrt{\nu}.$$

□

Let us fix now some reasonable values of parameters in the scheme (4.2.19). In the rest of this chapter we always assume that

$$\beta = \tfrac{1}{9}, \quad \gamma = \frac{\sqrt{\beta}}{1+\sqrt{\beta}} - \beta = \tfrac{5}{36}. \tag{4.2.22}$$

We have proved that it is possible to follow the central path, using the rule (4.2.19). Note that we can either increase or decrease the current value of t. The lower estimate for the rate of *increasing* t is

$$t_+ \geq \left(1 + \frac{5}{4+36\sqrt{\nu}}\right) \cdot t,$$

and the upper estimate for the rate of *decreasing* t is

$$t_+ \leq \left(1 - \frac{5}{4+36\sqrt{\nu}}\right) \cdot t.$$

Thus, the general scheme for solving the problem (4.2.12) is as follows.

Main path-following scheme
0. Set $t_0 = 0$. Choose an accuracy $\epsilon > 0$ and $x_0 \in \operatorname{dom} F$ such that $\parallel F'(x_0) \parallel_{x_0}^* \leq \beta.$
1. kth iteration ($k \geq 0$). Set $t_{k+1} = t_k + \frac{\gamma}{\parallel c \parallel_{x_k}^*}$, $x_{k+1} = x_k - [F''(x_k)]^{-1}(t_{k+1}c + F'(x_k)).$
2. Stop the process if $\epsilon t_k \geq \nu + \frac{(\beta+\sqrt{\nu})\beta}{1-\beta}$.

(4.2.23)

Let us give a complexity bound for the above scheme.

THEOREM 4.2.9 *The scheme (4.2.23) terminates no more than after N steps, where*

$$N \leq O\left(\sqrt{\nu} \ln \frac{\nu \parallel c \parallel_{x_F^*}^*}{\epsilon}\right).$$

Moreover, at the moment of termination we have $\langle c, x_N \rangle - c^* \leq \epsilon$.

Proof: Note that $r_0 \equiv \| x_0 - x_F^* \|_{x_0} \leq \frac{\beta}{1-\beta}$ (see Theorem 4.1.13). Therefore, in view of Theorem 4.1.6 we have

$$\frac{\gamma}{t_1} = \| c \|_{x_0}^* \leq \frac{1}{1-r_0} \| c \|_{x_F^*}^* \leq \frac{1-\beta}{1-2\beta} \| c \|_{x_F^*}^* .$$

Thus, $t_k \geq \frac{\gamma(1-2\beta)}{(1-\beta)\|c\|_{x_F^*}^*} \left(1 + \frac{\gamma}{\beta+\sqrt{\nu}}\right)^{k-1}$ for all $k \geq 1$. □

Let us discuss now the above complexity estimate. The main term in the complexity is

$$7.2\sqrt{\nu} \ln \frac{\nu \|c\|_{x_F^*}^*}{\epsilon}.$$

Note that the value $\nu \| c \|_{x_F^*}^*$ estimates the variation of the linear function $\langle c, x \rangle$ over the set Dom F (see Theorem 4.2.6). Thus, the ratio

$$\frac{\epsilon}{\nu \|c\|_{x_F^*}^*}$$

can be seen as a *relative accuracy* of the solution.

The process (4.2.23) has one serious drawback. Sometimes it is difficult to satisfy its starting condition

$$\| F'(x_0) \|_{x_0}^* \leq \beta.$$

In such cases we need an additional process for *finding* an appropriate starting point. We analyze the corresponding strategies in the next section.

4.2.5 Finding the analytic center

Thus, our goal now is to find an approximation to the *analytic center* of the set Dom F. Let us look at the following minimization problem:

$$\min\{F(x) \mid x \in \operatorname{dom} F\}, \tag{4.2.24}$$

where F is a ν-self-concordant barrier. In view of the needs of the previous section, we have to find an approximate solution $\bar{x} \in \operatorname{dom} F$ of this problem, which satisfies inequality

$$\| F'(\bar{x}) \|_{\bar{x}}^* \leq \beta,$$

for certain $\beta \in (0, 1)$.

In order to reach our goal, we can apply two different minimization schemes. The first one is a straightforward implementation of the damped Newton method. And the second one is based on path-following approach.

Consider the first scheme.

Damped Newton method for analytic centers
0. Choose $y_0 \in \operatorname{dom} F$.
1. kth iteration ($k \geq 0$). Set $$y_{k+1} = y_k - \frac{[F''(y_k)]^{-1}F'(y_k)}{1+\|F'(y_k)\|^*_{y_k}}.$$
2. Stop the process if $\| F'(y_k) \|^*_{y_k} \leq \beta$.

(4.2.25)

THEOREM 4.2.10 *The process (4.2.25) terminates no later than after* $\frac{1}{\omega(\beta)}(F(y_0) - F(x_F^*))$ *iterations.*

Proof: Indeed, in view of Theorem 4.1.12, we have

$$F(y_{k+1}) \leq F(y_k) - \omega(\lambda_F(y_k)) \leq F(y_k) - \omega(\beta).$$

Therefore $F(y_0) - k\,\omega(\beta) \geq F(y_k) \geq F(x_F^*)$. □

The implementation of the path-following approach is a little bit more complicated. Let us choose some $y_0 \in \operatorname{dom} F$. Define the *auxiliary central path* as follows:

$$y^*(t) = \arg \min_{y \in \operatorname{dom} F} [-t\langle F'(y_0), y\rangle + F(y)],$$

where $t \geq 0$. Note that this trajectory satisfies the equation

$$F'(y^*(t)) = tF'(y_0). \tag{4.2.26}$$

Therefore it connects two points, the starting point y_0 and the analytic center x_F^*:

$$y^*(1) = y_0, \quad y^*(0) = x_F^*.$$

We can follow this trajectory by the process (4.2.19) with *decreasing* t.

Let us estimate the rate of convergence of the auxiliary central path $y^*(t)$ to the analytic center.

LEMMA 4.2.2 *For any $t \geq 0$ we have*

$$\| F'(y^*(t)) \|^*_{y^*(t)} \leq (\nu + 2\sqrt{\nu}) \| F'(x_0) \|^*_{x^*_F} \cdot t.$$

Proof: This estimate follows from (4.2.26) and Corollary 4.2.1. □

Let us look now at the corresponding algorithmic scheme.

Auxiliary path-following scheme

0. Choose $y_0 \in \operatorname{Dom} F$. Set $t_0 = 1$.

1. kth iteration ($k \geq 0$). Set

$$\begin{aligned} t_{k+1} &= t_k - \frac{\gamma}{\| F'(y_0) \|^*_{y_k}}, \\ y_{k+1} &= y_k - [F''(y_k)]^{-1}(t_{k+1} F'(y_0) + F'(y_k)). \end{aligned} \tag{4.2.27}$$

2. Stop the process if $\| F'(y_k) \|_{y_k} \leq \frac{\sqrt{\beta}}{1+\sqrt{\beta}}$.
Set $\bar{x} = y_k - [F''(y_k)]^{-1} F'(y_k)$.

Note that the above scheme follows the auxiliary central path $y^*(t)$ as $t_k \to 0$. It updates the points $\{y_k\}$ satisfying the approximate centering condition

$$\| t_k F'(y_0) + F'(y_k) \|_{y_k} \leq \beta.$$

The termination criterion of this process,

$$\lambda_k = \| F'(y_k) \|_{y_k} \leq \frac{\sqrt{\beta}}{1+\sqrt{\beta}},$$

guarantees that $\| F'(\bar{x}) \|_{\bar{x}} \leq \left(\frac{\lambda_k}{1-\lambda_k} \right)^2 \leq \beta$ (see Theorem 4.1.14).

Let us derive a complexity estimate for this process.

THEOREM 4.2.11 *The process (4.2.27) terminates no later than after*

$$\frac{1}{\gamma}(\beta + \sqrt{\nu}) \ln \left[\frac{1}{\gamma}(\nu + 2\sqrt{\nu}) \| F'(x_0) \|^*_{x^*_F} \right]$$

iterations.

Proof: Recall that we have fixed the parameters:

$$\beta = \tfrac{1}{9}, \quad \gamma = \tfrac{\sqrt{\beta}}{1+\sqrt{\beta}} - \beta = \tfrac{5}{36}.$$

Note that $t_0 = 1$. Therefore, in view of Theorem 4.2.8 and Lemma 4.2.1, we have

$$t_{k+1} \leq \left(1 - \tfrac{\gamma}{\beta+\sqrt{\nu}}\right) t_k \leq \exp\left(-\tfrac{\gamma(k+1)}{\beta+\sqrt{\nu}}\right).$$

Further, in view of Lemma 4.2.2, we obtain

$$\begin{aligned} \| F'(y_k) \|^*_{y_k} &= \| (t_k F'(x_0) + F'(y_k)) - t_k F'(x_0) \|^*_{y_k} \\ &\leq \beta + t_k \| F'(x_0) \|^*_{y_k} \leq \beta + t_k(\nu + 2\sqrt{\nu}) \| F'(x_0) \|^*_{x^*_F}. \end{aligned}$$

Thus, the process is terminated at most when the following inequality holds:

$$t_k(\nu + 2\sqrt{\nu}) \| F'(x_0) \|^*_{x^*_F} \leq \tfrac{\sqrt{\beta}}{1+\sqrt{\beta}} - \beta = \gamma.$$

□

Now we can discuss the complexity of both schemes. The principal term in the complexity of the auxiliary path-following scheme is

$$7.2\sqrt{\nu}[\ln \nu + \ln \| F'(x_0) \|^*_{x^*_F}]$$

and for the auxiliary damped Newton method it is $O(F(y_0) - F(x^*_F))$. We cannot compare these estimates directly. However, a more sophisticated analysis demonstrates the advantages of the path-following approach. Note also that its complexity estimate naturally fits the complexity of the main path-following process. Indeed, if we apply (4.2.23) with (4.2.27), we get the following complexity bound for the whole process:

$$7.2\sqrt{\nu}\left[2\ln \nu + \ln \| F'(x_0) \|^*_{x^*_F} + \ln \| c \|^*_{x^*_F} + \ln \tfrac{1}{\epsilon}\right].$$

To conclude this section, note that for some problems it is difficult even to point out a starting point $y_0 \in \operatorname{dom} F$. In such cases we should apply one more auxiliary minimization process, which is similar to the process (4.2.27). We discuss this situation in the next section.

4.2.6 Problems with functional constraints

Let us consider the following minimization problem:

$$\begin{aligned} &\min\ f_0(x), \\ &\text{s.t}\ \ f_j(x) \leq 0,\ j = 1 \ldots m, \qquad (4.2.28) \\ &\quad\ \ \ x \in Q, \end{aligned}$$

where Q is a simple bounded closed convex set with nonempty interior and all functions f_j, $j = 0 \dots m$, are convex. We assume that the problem satisfies the Slater condition: There exists $\bar{x} \in \operatorname{int} Q$ such that $f_j(\bar{x}) < 0$ for all $j = 1 \dots m$.

Let us assume that we know an upper bound $\bar{\tau}$ such that $f_0(x) < \bar{\tau}$ for all $x \in Q$. Then, introducing two additional variables τ and κ, we can rewrite this problem in the standard form:

$$
\begin{aligned}
& \min \; \tau, \\
\text{s.t } \; & f_0(x) \le \tau, \\
& f_j(x) \le \kappa, \; j = 1 \dots m, \\
& x \in Q, \; \tau \le \bar{\tau}, \; \kappa \le 0.
\end{aligned}
\tag{4.2.29}
$$

Note that we can apply the interior-point methods to a problem only if we are able to construct the self-concordant barrier for the feasible set. In the current situation this means that we should be able to construct the following barriers:

- A self-concordant barrier $F_Q(x)$ for the set Q.
- A self-concordant barrier $F_0(x, \tau)$ for the epigraph of the objective function $f_0(x)$.
- Self-concordant barriers $F_j(x, \kappa)$ for the epigraphs of the functional constraints $f_j(x)$.

Let us assume that we can do that. Then the resulting self-concordant barrier for the feasible set of the problem (4.2.29) is as follows:

$$
\hat{F}(x, \tau, \kappa) = F_Q(x) + F_0(x, \tau) + \sum_{j=1}^{m} F_j(x, \kappa) - \ln(\bar{\tau} - \tau) - \ln(-\kappa).
$$

The parameter of this barrier is

$$
\hat{\nu} = \nu_Q + \nu_0 + \sum_{j=1}^{m} \nu_j + 2, \tag{4.2.30}
$$

where $\nu_{(\cdot)}$ are the parameters of the corresponding barriers.

Note that it could be still difficult to find a starting point from $\operatorname{dom} \hat{F}$. This domain is an intersection of the set Q with the epigraphs of the objective function and the constraints and with two additional constraints

$\tau \leq \bar{\tau}$ and $\kappa \leq 0$. If we have a point $x_0 \in \operatorname{int} Q$, then we can choose large enough τ_0 and κ_0 to guarantee

$$f_0(x_0) < \tau_0 < \bar{\tau}, \quad f_j(x_0) < \kappa_0, \; j = 1 \dots m,$$

but then the constraint $\kappa \leq 0$ could be violated.

In order to simplify our analysis, let us change notation. From now on we consider the problem

$$\begin{array}{rl} & \min \langle c, z \rangle, \\ \text{s.t.} & z \in S, \\ & \langle d, z \rangle \leq 0, \end{array} \tag{4.2.31}$$

where $z = (x, \tau, \kappa)$, $\langle c, z \rangle \equiv \tau$, $\langle d, z \rangle \equiv \kappa$ and S is the feasible set of the problem (4.2.29) without the constraint $\kappa \leq 0$. Note that we know a self-concordant barrier $F(z)$ for the set S and we can easily find a point $z_0 \in \operatorname{int} S$. Moreover, in view of our assumptions, the set

$$S(\alpha) = \{ z \in S \mid \langle d, z \rangle \leq \alpha \}$$

is bounded and it has nonempty interior for α large enough.

The process of solving the problem (4.2.31) consists of three stages.

1. Choose a starting point $z_0 \in \operatorname{int} S$ and an initial gap $\Delta > 0$. Set $\alpha = \langle d, z_0 \rangle + \Delta$. If $\alpha \leq 0$, then we can use the two-stage process described in Section 4.2.5. Otherwise we do the following. First, we find an approximate analytic center of the set $S(\alpha)$, generated by the barrier

$$\tilde{F}(z) = F(z) - \ln(\alpha - \langle d, z \rangle).$$

Namely, we find a point $\tilde{z}$ satisfying the condition

$$\lambda_{\tilde{F}}(\tilde{z}) \equiv \langle \tilde{F}''(\tilde{z})^{-1} \left(F'(\tilde{z}) + \tfrac{d}{\alpha - \langle d, \tilde{z} \rangle} \right), F'(\tilde{z}) + \tfrac{d}{\alpha - \langle d, \tilde{z} \rangle} \rangle^{1/2} \leq \beta.$$

In order to generate such a point, we can use the auxiliary schemes discussed in Section 4.2.5.

2. The next stage consists in following the central path $z(t)$ defined by the equation

$$td + \tilde{F}'(z(t)) = 0, \quad t \geq 0.$$

Note that the previous stage provides us with a reasonable approximation to the analytic center $z(0)$. Therefore we can follow this path, using the process (4.2.19). This trajectory leads us to the solution of the minimization problem

$$\min\{ \langle d, z \rangle \mid z \in S(\alpha) \}.$$

In view of the Slater condition for problem (4.2.31), the optimal value of this problem is strictly negative.

The goal of this stage consists in finding an approximation to the analytic center of the set

$$\bar{S} = \{z \in S(\alpha) \mid \langle d, z\rangle \leq 0\},$$

generated by the barrier

$$\bar{F}(z) = \tilde{F}(z) - \ln(-\langle d, z\rangle).$$

This point, z_*, satisfies the equation

$$\tilde{F}'(z_*) - \frac{d}{\langle d, z_*\rangle} = 0.$$

Therefore z^* is a point of the central path $z(t)$. The corresponding value of the penalty parameter t_* is

$$t_* = -\frac{1}{\langle d, z_*\rangle} > 0.$$

This stage ends up with a point $\bar{z}$, satisfying the condition

$$\lambda_{\bar{F}}(\bar{z}) \equiv \langle \tilde{F}''(\bar{z})^{-1}\left(\tilde{F}'(\bar{z}) - \frac{d}{\langle d, \bar{z}\rangle}\right), \tilde{F}'(\bar{z}) - \frac{d}{\langle d, \bar{z}\rangle}\rangle^{1/2} \leq \beta.$$

3. Note that $\bar{F}''(z) > \tilde{F}''(z)$. Therefore, the point $\bar{z}$, computed at the previous stage satisfies inequality

$$\lambda_{\bar{F}}(\bar{z}) \equiv \langle \bar{F}''(\bar{z})^{-1}\left(\tilde{F}'(\bar{z}) - \frac{d}{\langle d, \bar{z}\rangle}\right), \tilde{F}'(\bar{z}) - \frac{d}{\langle d, \bar{z}\rangle}\rangle^{1/2} \leq \beta.$$

This means that we have a good approximation of the analytic center of the set $\bar{S}$ and we can apply the main path-following scheme (4.2.23) to solve the problem

$$\min\{\langle c, z\rangle \mid z \in \bar{S}\}.$$

Clearly, this problem is equivalent to (4.2.31).

We omit the detailed complexity analysis of the above three-stage scheme. It could be done similarly to the analysis of Section 4.2.5. The main term in the complexity of this scheme is proportional to the product of $\sqrt{\hat{\nu}}$ (see (4.2.30)) and the sum of the logarithm of desired accuracy ϵ and the logarithms of some structural characteristics of the problem (size of the region, deepness of Slater condition, etc.).

Thus, we have shown that we can apply efficient interior point methods to all problems, for which we can point out some self-concordant barriers for the basic feasible set Q and for the epigraphs of functional constraints. Our main goal now is to describe the classes of convex problems, for which such barriers can be constructed in a computable form. Note that we have an exact characteristic of the quality of the self-concordant barrier. That is the value of its parameter: The smaller it is, the more efficient will be the corresponding path-following scheme. In the next section we discuss our possibilities in applying the developed theory to particular convex problems.

4.3 Applications of structural optimization

(Bounds on parameters of self-concordant barriers; Linear and quadratic optimization; Semidefinite optimization; Extremal ellipsoids; Separable problems; Geometric optimization; Approximation in l_p norms; Choice of optimization scheme.)

4.3.1 Bounds on parameters of self-concordant barriers

In the previous section we have discussed a path-following scheme for solving the following problem:

$$\min_{x \in Q} \langle c, x \rangle, \qquad (4.3.1)$$

where Q is a closed convex set with nonempty interior, for which we know a ν-self-concordant barrier $F(x)$. Using such a barrier, we can solve (4.3.1) in $O\left(\sqrt{\nu} \cdot \ln \frac{\nu}{\epsilon}\right)$ iterations of a path-following scheme. Recall that the most difficult part of each iteration is the solution of a system of linear equations.

In this section we study the limits of applicability of this approach. We discuss the lower and upper bounds for the parameters of self-concordant barriers; we also discuss some classes of convex problems, for which the model (4.3.1) can be created in a computable form.

Let us start from lower bounds on barrier parameters.

LEMMA 4.3.1 *Let $f(t)$ be a ν-self-concordant barrier for the interval $(\alpha, \beta) \subset R^1$, $\alpha < \beta < \infty$. Then*

$$\nu \geq \kappa \equiv \sup_{t \in (\alpha,\beta)} \frac{(f'(t))^2}{f''(t)} \geq 1.$$

Proof: Note that $\nu \geq \kappa$ by definition. Let us assume that $\kappa < 1$. Since $f(t)$ is a barrier for (α, β), there exists a value $\bar{\alpha} \in (\alpha, \beta)$ such that $f'(t) > 0$ for all $t \in [\bar{\alpha}, \beta)$.

Consider the function $\phi(t) = \frac{(f'(t))^2}{f''(t)}$, $t \in [\bar{\alpha}, \beta)$. Then, since $f'(t) > 0$, $f(t)$ is self-concordant and $\phi(t) \leq \kappa < 1$, we have

$$\begin{aligned}
\phi'(t) &= 2f'(t) - \left(\frac{f'(t)}{f''(t)}\right)^2 f'''(t) \\
&= f'(t)\left(2 - \frac{f'(t)}{\sqrt{f''(t)}} \cdot \frac{f'''(t)}{[f''(t)]^{3/2}}\right) \geq 2(1 - \sqrt{\kappa})f'(t).
\end{aligned}$$

Hence, for all $t \in [\bar{\alpha}, \beta)$ we obtain $\phi(t) \geq \phi(\bar{\alpha}) + 2(1 - \sqrt{\kappa})(f(t) - f(\bar{\alpha}))$. This is a contradiction since $f(t)$ is a barrier and $\phi(t)$ is bounded from above. □

COROLLARY 4.3.1 *Let $F(x)$ be a ν-self-concordant barrier for $Q \subset R^n$. Then $\nu \geq 1$.*

Proof: Indeed, let $x \in \operatorname{int} Q$. Since $Q \subset R^n$, there exists a nonzero direction $u \in R^n$ such that the line $\{y = x + tu, \ t \in R^1\}$ intersects the boundary of the set Q. Therefore, considering the function $f(t) = F(x + tu)$, and using Lemma 4.3.1, we get the result. □

Let us prove a simple lower bound for parameters of self-concordant barriers for unbounded sets.

Let Q be a closed convex set with nonempty interior. Consider $\bar{x} \in \operatorname{int} Q$. Assume that there exists a nontrivial set of *recession* directions $\{p_1, \dots, p_k\}$ of the set Q:

$$\bar{x} + \alpha p_i \in Q \quad \forall \alpha \geq 0.$$

THEOREM 4.3.1 *Let positive coefficients $\{\beta_i\}_{i=1}^k$ satisfy condition*

$$\bar{x} - \beta_i\, p_i \notin \operatorname{int} Q, \quad i = 1 \dots k.$$

If for some positive $\alpha_1, \dots, \alpha_k$ we have $\bar{y} = \bar{x} - \sum\limits_{i=1}^k \alpha_i p_i \in Q$, then the parameter ν of any self-concordant barrier for Q satisfies inequality:

$$\nu \geq \sum_{i=1}^k \frac{\alpha_i}{\beta_i}.$$

Proof: Let $F(x)$ be a ν-self-concordant barrier for the set Q. Since p_i is a recession direction, we have

$$\langle F'(\bar{x}), -p_i \rangle \geq \langle F''(\bar{x}) p_i, p_i \rangle^{1/2} \equiv \| p_i \|_{\bar{x}},$$

(since otherwise the function $f(t) = F(\bar{x} + tp)$ attains its minimum; see Theorem 4.1.11).

Note that $\bar{x} - \beta_i\, p_i \notin Q$. Therefore, in view of Theorem 4.1.5, the norm of the direction p_i is large enough: $\beta_i \parallel p_i \parallel_{\bar{x}} \geq 1$. Hence, in view of Theorem 4.2.4, we obtain

$$\nu \geq \langle F'(\bar{x}), \bar{y} - \bar{x} \rangle = \langle F'(\bar{x}), -\sum_{i=1}^{k} \alpha_i p_i \rangle \geq \sum_{i=1}^{k} \alpha_i \parallel p_i \parallel_{\bar{x}} \geq \sum_{i=1}^{k} \frac{\alpha_i}{\beta_i}.$$

□

Let us present now an existence theorem for self-concordant barriers. Consider a closed convex set Q, $\operatorname{int} Q \neq \emptyset$, and assume that Q contains no straight line. Define a *polar set* of Q with respect to some point $\bar{x} \in \operatorname{int} Q$:

$$P(\bar{x}) = \{s \in R^n \mid \langle s, x - \bar{x} \rangle \leq 1, \quad \forall x \in Q\}.$$

It can be proved that for any $x \in \operatorname{int} Q$ the set $P(x)$ is a bounded closed convex set with nonempty interior. Denote $V(x) = \operatorname{vol}_n P(x)$.

THEOREM 4.3.2 *There exist absolute constants c_1 and c_2, such that the function*

$$U(x) = c_1 \cdot \ln V(x)$$

is a $(c_2 \cdot n)$-self-concordant barrier for Q. □

Function $U(x)$ is called the *universal barrier* for the set Q. Note that the analytical complexity of problem (4.3.1), equipped with a universal barrier, is $O\left(\sqrt{n} \cdot \ln \frac{n}{\epsilon}\right)$. Recall that such efficiency estimate is *impossible*, if we use a local black-box oracle (see Theorem 3.2.5).

The above result has mainly a theoretical interest. In general, the universal barrier $U(x)$ cannot be easily computed. However, Theorem 4.3.2 demonstrates that such barriers, in principle, can be found for *any* convex set. Thus, the applicability of our approach is restricted only by abilities of constructing a *computable* self-concordant barrier, hopefully with a small value of the parameter. The process of creating the *barrier model* of the initial problem, can be hardly described in a formal way. For each particular problem there could be many different barrier models, and we should choose the best one, taking into account the value of the parameter of the self-concordant barrier, the complexity of its gradient and Hessian, and the complexity of solution of the Newton system. In the rest of this section we will see how that can be done for some *standard* problem classes of convex optimization.

4.3.2 Linear and quadratic optimization

Let us start from linear optimization problem:

$$
\begin{aligned}
&\min_{x \in R^n} \langle c, x \rangle, \\
\text{s.t} \quad & Ax = b, \qquad\qquad (4.3.2) \\
& x^{(i)} \geq 0, \ i = 1 \dots n, \quad (\Leftrightarrow x \in R^n_+)
\end{aligned}
$$

where A is an $(m \times n)$-matrix, $m < n$. The inequalities in this problem define the *positive orthant* in R^n. This set can be equipped with the following self-concordant barrier:

$$F(x) = -\sum_{i=1}^{n} \ln x^{(i)}, \qquad \nu = n,$$

(see Example 4.2.1 and Theorem 4.2.2). This barrier is called the *standard logarithmic barrier* for R^n_+.

In order to solve the problem (4.3.2), we have to use a restriction of the barrier $F(x)$ onto affine subspace $\{x : \ Ax = b\}$. Since this restriction is an n-self-concordant barrier (see Theorem 4.2.3), the complexity estimate for the problem (4.3.2) is $O\left(\sqrt{n} \cdot \ln \frac{n}{\epsilon}\right)$ iterations of a path-following scheme.

Let us prove that the standard logarithmic barrier is optimal for R^n_+.

LEMMA 4.3.2 *Parameter ν of any self-concordant barrier for R^n_+ satisfies the inequality $\nu \geq n$.*

Proof: Let us choose

$$
\begin{aligned}
\bar{x} &= e \equiv (1, \dots, 1)^T \in \operatorname{int} R^n_+, \\
p_i &= e_i, \quad i = 1 \dots n,
\end{aligned}
$$

where e_i is the ith coordinate vector of R^n. Clearly, the conditions of Theorem 4.3.1 are satisfied with $\alpha_i = \beta_i = 1$, $i = 1 \dots n$. Therefore

$$\nu \geq \sum_{i=1}^{n} \frac{\alpha_i}{\beta_i} = n.$$

□

Note that the above lower bound is valid only for the entire set R^n_+. The lower bound for intersection $\{x \in R^n_+ \mid Ax = b\}$ can be smaller.

Let us look now at a quadratically constrained quadratic optimization problem:

$$\begin{aligned} \min_{x \in R^n} \; & q_0(x) = \alpha_0 + \langle a_0, x\rangle + \tfrac{1}{2}\langle A_0 x, x\rangle, \\ \text{s.t} \; & q_i(x) = \alpha_i + \langle a_i, x\rangle + \tfrac{1}{2}\langle A_i x, x\rangle \le \beta_i, \; i = 1\dots m, \end{aligned} \tag{4.3.3}$$

where A_i are some positive semidefinite $(n \times n)$-matrices. Let us rewrite this problem in a standard form:

$$\begin{aligned} \min_{x,\tau} \; & \tau, \\ \text{s.t} \; & q_0(x) \le \tau, \\ & q_i(x) \le \beta_i, \; i = 1\dots m, \\ & x \in R^n, \; \tau \in R^1. \end{aligned} \tag{4.3.4}$$

The feasible set of this problem can be equipped with the following self-concordant barrier:

$$F(x,\tau) = -\ln(\tau - q_0(x)) - \sum_{i=1}^{m} \ln(\beta_i - q_i(x)), \quad \nu = m + 1,$$

(see Example 4.2.1, and Theorem 4.2.2). Thus, the complexity bound for problem (4.3.3) is $O\left(\sqrt{m+1} \cdot \ln \frac{m}{\epsilon}\right)$ iterations of a path-following scheme. Note this estimate *does not depend* on n.

In many applications the functional components of the problem include a nonsmooth quadratic term of the form $\| Ax - b \|$. Let us show that we can treat such terms using interior-point technique.

LEMMA 4.3.3 *The function*

$$F(x,t) = -\ln(t^2 - \| x \|^2)$$

is a 2-self-concordant barrier for the convex set [5]

$$K_2 = \{(x,t) \in R^{n+1} \mid t \ge \| x \|\}.$$

Proof: Let us fix a point $z = (x,t) \in \operatorname{int} K_2$ and a nonzero direction $u = (h,\tau) \in R^{n+1}$. Denote $\xi(\alpha) = (t + \alpha\tau)^2 - \| x + \alpha h \|^2$. We need to compare the derivatives of function

$$\phi(\alpha) = F(z + \alpha u) = -\ln \xi(\alpha)$$

[5] Depending on the field, this set has different names: Lorentz cone, ice-cream cone, second-order cone.

at $\alpha = 0$. Denote $\phi^{(\cdot)} = \phi^{(\cdot)}(0)$, $\xi^{(\cdot)} = \xi^{(\cdot)}(0)$. Then

$$\xi' = 2(t\tau - \langle x, h\rangle), \quad \xi'' = 2(\tau^2 - \| h \|^2),$$

$$\phi' = -\frac{\xi'}{\xi}, \quad \phi'' = \left(\frac{\xi'}{\xi}\right)^2 - \frac{\xi''}{\xi}, \quad \phi''' = 3\frac{\xi'\xi''}{\xi^2} - 2\left(\frac{\xi'}{\xi}\right)^3.$$

Note the inequality $2\phi'' \geq (\phi')^2$ is equivalent to $(\xi')^2 \geq 2\xi\xi''$. Thus, we need to prove that for any (h, τ) we have

$$(t\tau - \langle x, h\rangle)^2 \geq (t^2 - \| x \|^2)(\tau^2 - \| h \|^2). \tag{4.3.5}$$

Clearly, we can restrict ourselves by $| \tau | > \| h \|$ (otherwise the right-hand side of the above inequality is nonpositive). Moreover, in order to minimize the left-hand side, we should chose sign τ = sign $\langle x, h\rangle$ (thus, let $\tau > 0$), and $\langle x, h\rangle = \| x \| \cdot \| h \|$. Substituting these values in (4.3.5), we get a valid inequality.

Finally, since $0 \leq \frac{\xi\xi''}{(\xi')^2} \leq \frac{1}{2}$ and $[1 - \xi]^{3/2} \geq 1 - \frac{3}{2}\xi$, we get the following:

$$\frac{|\phi'''|}{(\phi'')^{3/2}} = 2\frac{|\xi'| \cdot |(\xi')^2 - \frac{3}{2}\xi\xi''|}{[(\xi')^2 - \xi\xi'']^{3/2}} \leq 2.$$

□

Let us prove that the barrier described in the above statement is optimal for the second-order cone.

LEMMA 4.3.4 *Parameter ν of any self-concordant barrier for the set K_2 satisfies inequality $\nu \geq 2$.*

Proof: Let us choose $\bar{z} = (0, 1) \in \operatorname{int} K_2$ and some $h \in R^n$, $\| h \| = 1$. Define

$$p_1 = (h, 1), \quad p_2 = (-h, 1), \quad \alpha_1 = \alpha_2 = \frac{1}{2}, \quad \beta_1 = \beta_2 = \frac{1}{2}.$$

Note that for all $\gamma \geq 0$ we have $\bar{z} + \gamma p_i = (\pm\gamma h, 1 + \gamma) \in K_2$ and

$$\bar{z} - \beta_i p_i = (\pm\tfrac{1}{2}h, \tfrac{1}{2}) \notin \operatorname{int} K_2,$$

$$\bar{z} - \alpha_1 p_1 - \alpha_2 p_2 = (-\tfrac{1}{2}h + \tfrac{1}{2}h, 1 - \tfrac{1}{2} - \tfrac{1}{2}) = 0 \in K_2.$$

Therefore, the conditions of Theorem 4.3.1 are satisfied and

$$\nu \geq \frac{\alpha_1}{\beta_1} + \frac{\alpha_2}{\beta_2} = 2.$$

□

4.3.3 Semidefinite optimization

In semidefinite optimization the decision variables are some matrices. Let $X = \{X^{(i,j)}\}_{i,j=1}^n$ be a symmetric $n \times n$-matrix (notation: $X \in S^{n\times n}$). The linear space $S^{n\times n}$ can be provided with the following inner product: for any $X, Y \in S^{n\times n}$ define

$$\langle X, Y\rangle_F = \sum_{i=1}^n \sum_{j=1}^n X^{(i,j)} Y^{(i,j)}, \quad \| X \|_F = \langle X, X\rangle_F^{1/2}.$$

Sometimes the value $\| X \|_F$ is called the *Frobenius norm* of matrix X. For symmetric matrices X and Y we have the following identity:

$$\begin{aligned}
&\langle X, Y\cdot Y\rangle_F \\
&= \sum_{i=1}^n \sum_{j=1}^n X^{(i,j)} \sum_{k=1}^n Y^{(i,k)} Y^{(j,k)} = \sum_{i=1}^n \sum_{j=1}^n \sum_{k=1}^n X^{(i,j)} Y^{(i,k)} Y^{(j,k)} \\
&= \sum_{k=1}^n \sum_{j=1}^n Y^{(k,j)} \sum_{i=1}^n X^{(j,i)} Y^{(i,k)} = \sum_{k=1}^n \sum_{j=1}^n Y^{(k,j)} (XY)^{(j,k)} \\
&= \sum_{k=1}^n (YXY)^{(k,k)} = \mathrm{Trace}\,(YXY) = \langle YXY, I_n\rangle_F.
\end{aligned} \tag{4.3.6}$$

In semidefinite optimization problems a nontrivial part of constraints is formed by the *cone of positive semidefinite* $n\times n$-matrices $\mathcal{P}_n \subset S^{n\times n}$. Recall that $X \in \mathcal{P}_n$ if and only if $\langle Xu, u\rangle \geq 0$ for any $u \in R^n$. If $\langle Xu, u\rangle > 0$ for all nonzero u, we call X *positive definite*. Such matrices form an interior of cone $\mathcal{P}_n$. Note that $\mathcal{P}_n$ is a closed convex set.

The general formulation of the semidefinite optimization problem is as follows:

$$\begin{aligned}
&\min \langle C, X\rangle_F, \\
&\text{s.t } \langle A_i, X\rangle_F = b_i, \; i = 1 \ldots m, \\
&X \in \mathcal{P}_n,
\end{aligned} \tag{4.3.7}$$

where C and A_i belong to $S^{n\times n}$. In order to apply a path-following scheme to this problem, we need a self-concordant barrier for $\mathcal{P}_n$.

Let matrix X belong to $\mathrm{int}\,\mathcal{P}_n$. Denote $F(X) = -\ln\det X$. Clearly

$$F(X) = -\ln \prod_{i=1}^n \lambda_i(X),$$

where $\{\lambda_i(X)\}_{i=1}^n$ is the set of eigenvalues of matrix X.

LEMMA 4.3.5 *Function $F(X)$ is convex and $F'(X) = -X^{-1}$. For any direction $\Delta \in S^{n\times n}$ we have*

$$\begin{aligned}\langle F''(X)\Delta, \Delta\rangle_F &= \| X^{-1/2}\Delta X^{-1/2} \|_F^2 = \langle X^{-1}\Delta X^{-1}, \Delta\rangle_F \\ &= \text{Trace}\left([X^{-1/2}\Delta X^{-1/2}]^2\right), \\ D^3F(x)[\Delta,\Delta,\Delta] &= -2\langle I_n, [X^{-1/2}\Delta X^{-1/2}]^3\rangle_F \\ &= -2\text{Trace}\left([X^{-1/2}\Delta X^{-1/2}]^3\right).\end{aligned}$$

Proof: Let us fix some $\Delta \in S^{n\times n}$ and $X \in \operatorname{int} \mathcal{P}_n$ such that $X+\Delta \in \mathcal{P}_n$. Then

$$\begin{aligned}F(X+\Delta) - F(X) &= -\ln\det(X+\Delta) - \ln\det X \\ &= -\ln\det(I_n + X^{-1/2}\Delta X^{-1/2}) \\ &\geq -\ln\left[\tfrac{1}{n}\text{Trace}\,(I_n + X^{-1/2}\Delta X^{-1/2})\right]^n \\ &= -n\ln\left[1 + \tfrac{1}{n}\langle I_n, X^{-1/2}\Delta X^{-1/2}\rangle_F\right] \\ &\geq -\langle I_n, X^{-1/2}\Delta X^{-1/2}\rangle_F = -\langle X^{-1}, \Delta\rangle_F.\end{aligned}$$

Thus, $-X^{-1} \in \partial F(X)$. Therefore F is convex (Lemma 3.1.6) and $F'(x) = -X^{-1}$ (Lemma 3.1.7).

Further, consider function $\phi(\alpha) \equiv \langle F'(X+\alpha\Delta), \Delta\rangle_F$, $\alpha \in [0,1]$. Then

$$\begin{aligned}\phi(\alpha) - \phi(0) &= \langle X^{-1} - (X+\alpha\Delta)^{-1}, \Delta\rangle_F \\ &= \langle (X+\alpha\Delta)^{-1}[(X+\alpha\Delta) - X]X^{-1}, \Delta\rangle_F \\ &= \alpha\langle (X+\alpha\Delta)^{-1}\Delta X^{-1}, \Delta\rangle_F.\end{aligned}$$

Thus, $\phi'(0) = \langle F''(X)\Delta, \Delta\rangle_F = \langle X^{-1}\Delta X^{-1}, \Delta\rangle_F$.

The last expression can be proved in a similar way by differentiating the function $\psi(\alpha) = \langle (X+\alpha\Delta)^{-1}\Delta(X+\alpha\Delta)^{-1}, \Delta\rangle_F$. □

THEOREM 4.3.3 *Function $F(X)$ is an n-self-concordant barrier for $\mathcal{P}_n$.*

Proof: Let us fix $X \in \operatorname{int} \mathcal{P}_n$ and $\Delta \in S^{n\times n}$. Denote $Q = X^{-1/2}\Delta X^{-1/2}$ and $\lambda_i = \lambda_i(Q)$, $i = 1 \dots n$. Then, in view of Lemma 4.3.5 we have

$$\begin{aligned}
\langle F'(X), \Delta\rangle_F &= \sum_{i=1}^{n} \lambda_i, \\
\langle F''(X)\Delta, \Delta\rangle_F &= \sum_{i=1}^{n} \lambda_i^2, \\
D^3F(X)[\Delta,\Delta,\Delta] &= -2\sum_{i=1}^{n} \lambda_i^3.
\end{aligned}$$

Using two standard inequalities

$$\left(\sum_{i=1}^{n} \lambda_i\right)^2 \le n\sum_{i=1}^{n} \lambda_i^2, \quad \left|\sum_{i=1}^{n} \lambda_i^3\right| \le \left(\sum_{i=1}^{n} \lambda_i^2\right)^{3/2},$$

we obtain

$$\begin{aligned}
\langle F'(X), \Delta\rangle_F^2 &\le n\langle F''(X)\Delta, \Delta\rangle_F, \\
\mid D^3F(X)[\Delta,\Delta,\Delta] \mid &\le 2\langle F''(X)\Delta, \Delta\rangle_F^{3/2}.
\end{aligned}$$

□

Let us prove that $F(X) = -\ln\det X$ is the optimal barrier for $\mathcal{P}_n$.

LEMMA 4.3.6 *Parameter ν of any self-concordant barrier for cone $\mathcal{P}_n$ satisfies inequality $\nu \ge n$.*

Proof: Let us choose $\bar{X} = I_n \in \operatorname{int} \mathcal{P}_n$ and the directions $P_i = e_i e_i^T$, $i = 1 \dots n$, where e_i is the ith coordinate vector of R^n. Note that for any $\gamma \ge 0$ we have $I_n + \gamma P_i \in \operatorname{int} \mathcal{P}_n$. Moreover,

$$I_n - e_i e_i^T \notin \operatorname{int} \mathcal{P}_n, \quad I_n - \sum_{i=1}^{n} e_i e_i^T = 0 \in \mathcal{P}_n.$$

Therefore, the conditions of Theorem 4.3.1 are satisfied with $\alpha_i = \beta_i = 1$, $i = 1 \dots n$, and we obtain $\nu \ge \sum_{i=1}^{n} \frac{\alpha_i}{\beta_i} = n$. □

As in the linear optimization problem (4.3.2), in problem (4.3.7) we need to use restriction of $F(X)$ onto the set

$$\mathcal{L} = \{X : \langle A_i, X\rangle_F = b_i, \ i = 1 \dots m\}.$$

This restriction is an n-self-concordant barrier in view of Theorem 4.2.3. Thus, the complexity bound of the problem (4.3.7) is $O\left(\sqrt{n}\cdot \ln\frac{n}{\epsilon}\right)$ iterations of a path-following scheme. Note that this estimate is very encouraging since the dimension of the problem (4.3.7) is $\frac{1}{2}n(n+1)$.

Let us estimate the arithmetical cost of each iteration of a path-following scheme (4.2.23) as applied to the problem (4.3.7). Note that we work with a restriction of the barrier $F(X)$ onto the set $\mathcal{L}$. In view of Lemma 4.3.5, each Newton step consists in solving the following problem:

$$\min_{\Delta}\{\langle U, \Delta\rangle_F + \tfrac{1}{2}\langle X^{-1}\Delta X^{-1}, \Delta\rangle_F : \langle A_i, \Delta\rangle_F = 0, \; i = 1\dots m\},$$

where $X \succ 0$ belongs to $\mathcal{L}$ and U is a combination of the cost matrix C and the gradient $F'(X)$. In accordance to Corollary 1.2.1, the solution of this problem can be found from the following system of linear equations:

$$\begin{aligned} U + X^{-1}\Delta X^{-1} &= \sum_{j=1}^{m} \lambda^{(j)} A_j, \\ \langle A_i, \Delta\rangle_F &= 0, \quad i = 1\dots m. \end{aligned} \tag{4.3.8}$$

From the first equation in (4.3.8) we get

$$\Delta = X\left[-U + \sum_{j=1}^{m} \lambda^{(j)} A_j\right] X. \tag{4.3.9}$$

Substituting this expression in the second equation in (4.3.8), we get the linear system

$$\sum_{j=1}^{m} \lambda^{(j)} \langle A_i, X A_j X\rangle_F = \langle A_i, XUX\rangle_F, \quad i = 1\dots m, \tag{4.3.10}$$

which can be written in a matrix form as $S\lambda = d$ with

$$S^{(i,j)} = \langle A_i, X A_j X\rangle_F, \quad d^{(j)} = \langle U, X A_j X\rangle_F, \quad i, j = 1\dots n.$$

Thus, a straightforward strategy of solving the system (4.3.8) consists in the following steps.

- Compute matrices XA_jX, $j = 1\dots m$. Cost: $O(mn^3)$ operations.
- Compute the elements of S and d. Cost: $O(m^2n^2)$ operations.
- Compute $\lambda = S^{-1}d$. Cost: $O(m^3)$ operations.
- Compute Δ by (4.3.9). Cost: $O(mn^2)$ operations.

Taking into account that $m \le \frac{n(n+1)}{2}$ we conclude that the complexity of one Newton step does not exceed

$$\boxed{O(n^2(m+n)m) \text{ arithmetic operations.}} \tag{4.3.11}$$

However, if the matrices A_j possess a certain structure, then this estimate can be significantly improved. For example, if all A_j are of rank 1:

$$A_j = a_j a_j^T, \quad a_j \in R^n, \quad j = 1 \dots m,$$

then the computation of the Newton step can be done in

$$\boxed{O((m+n)^3) \text{ arithmetic operations.}} \tag{4.3.12}$$

We leave the justification of this claim as an exercise for the reader.

To conclude this section, note that in many important applications we can use the barrier $-\ln\det(\cdot)$ for treating some functions of eigenvalues. Consider, for example, a matrix $\mathcal{A}(x) \in S^{n\times n}$, which depends linearly on x. Then the convex region

$$\{(x,t) \mid \max_{1\le i\le n} \lambda_i(\mathcal{A}(x)) \le t\},$$

can be described by a self-concordant barrier

$$F(x,t) = -\ln\det(tI_n - \mathcal{A}(x)).$$

The value of the parameter of this barrier is equal to n.

4.3.4 Extremal ellipsoids

In some applications we are interested in approximating polytopes by ellipsoids. Let us consider the most important examples.

4.3.4.1 Circumscribed ellipsoid

> Given by a set of points $a_1, \dots, a_m \in R^n$, find an ellipsoid W, which contains all points $\{a_i\}$ and which volume is as small as possible.

Let us pose this problem in a formal way. First of all note, that any bounded ellipsoid $W \subset R^n$ can be represented as

$$W = \{x \in R^n \mid x = H^{-1}(v+u), \; \| u \| \le 1\},$$

where $H \in \operatorname{int} \mathcal{P}_n$ and $v \in R^n$. Then inclusion $a \in W$ is equivalent to inequality $\| Ha - v \| \le 1$. Note also that

$$\operatorname{vol}_n W = \operatorname{vol}_n B_2(0,1) \cdot \det H^{-1} = \tfrac{\operatorname{vol}_n B_2(0,1)}{\det H}.$$

Thus, our problem is as follows:

$$\begin{aligned} &\min_{H,v,\tau} \ \tau, \\ \text{s.t.} \quad & -\ln\det H \leq \tau, \\ & \| Ha_i - v \| \leq 1, \ i = 1 \ldots m, \\ & H \in \mathcal{P}_n, \ v \in R^n, \tau \in R^1. \end{aligned} \tag{4.3.13}$$

In order to solve this problem by an interior-point scheme we need to find a self-concordant barrier for a feasible set. At this moment we know such barriers for all components of this problem except the first inequality.

LEMMA 4.3.7 *Function*

$$-\ln\det H - \ln(\tau + \ln\det H)$$

is an $(n+1)$*-self-concordant barrier for the set*

$$\{(H,\tau) \in S^{n\times n} \times R^1 \mid \tau \geq -\ln\det H, \ H \in \mathcal{P}_n\}.$$

□

Thus, we can use the following barrier:

$$F(H,v,\tau) = -\ln\det H - \ln(\tau + \ln\det H) - \sum_{i=1}^{m} \ln(1 - \| Ha_i - v \|^2),$$

$$\nu = m + n + 1.$$

The corresponding complexity bound is $O\left(\sqrt{m+n+1} \cdot \ln \frac{m+n}{\epsilon}\right)$ iterations of a path-following scheme.

4.3.4.2 Inscribed ellipsoid with fixed center

> Let Q be a convex polytope defined by a set of linear inequalities:
>
> $$Q = \{x \in R^n \mid \langle a_i, x\rangle \leq b_i, \ i = 1 \ldots m\},$$
>
> and let $v \in \operatorname{int} Q$. Find an ellipsoid W, centered at v, such that $W \subset Q$ and which volume is as big as possible.

Let us fix some $H \in \operatorname{int} \mathcal{P}_n$. We can represent the ellipsoid W as

$$W = \{x \in R^n \mid \langle H^{-1}(x - v), x - v\rangle \leq 1\}.$$

We need the following simple result.

LEMMA 4.3.8 *Let $\langle a, v\rangle < b$. Inequality $\langle a, x\rangle \leq b$ is valid for any $x \in W$ if and only if*

$$\langle Ha, a\rangle \leq (b - \langle a, v\rangle)^2.$$

Proof: In view of Lemma 3.1.12, we have

$$\max_u \{\langle a, u\rangle \mid \langle H^{-1}u, u\rangle \leq 1\} = \langle Ha, a\rangle^{1/2}.$$

Therefore we need to ensure

$$\begin{aligned}
\max_{x \in W} \langle a, x\rangle &= \max_{x \in W}[\langle a, x - v\rangle + \langle a, v\rangle] \\
&= \langle a, v\rangle + \max_x \{\langle a, u\rangle \mid \langle H^{-1}u, u\rangle \leq 1\} \\
&= \langle a, v\rangle + \langle Ha, a\rangle^{1/2} \leq b.
\end{aligned}$$

This proves our statement since $\langle a, v\rangle < b$. □

Note that $\mathrm{vol}_n W = \mathrm{vol}_n B_2(0,1)[\det H]^{1/2}$. Hence, our problem is as follows:

$$\begin{aligned}
&\min_{H,\tau} \tau, \\
\text{s.t.} \quad & -\ln\det H \leq \tau, \\
& \langle Ha_i, a_i\rangle \leq (b_i - \langle a_i, v\rangle)^2, \ i = 1 \dots m, \\
& H \in \mathcal{P}_n, \ \tau \in R^1.
\end{aligned} \tag{4.3.14}$$

In view of Lemma 4.3.7, we can use the following self-concordant barrier:

$$\begin{aligned}
F(H, \tau) &= -\ln\det H - \ln(\tau + \ln\det H) \\
&\quad - \sum_{i=1}^m \ln[(b_i - \langle a_i, v\rangle)^2 - \langle Ha_i, a_i\rangle], \\
\nu &= m + n + 1.
\end{aligned}$$

The efficiency estimate of the corresponding path-following scheme is $O\left(\sqrt{m+n+1} \cdot \ln \frac{m+n}{\epsilon}\right)$ iterations.

4.3.4.3 Inscribed ellipsoid with free center

Let Q be a convex polytope defined by a set of linear inequalities:

$$Q = \{x \in R^n \mid \langle a_i, x\rangle \leq b_i, \ i = 1 \dots m\},$$

and let $\operatorname{int} Q \neq \emptyset$. Find an ellipsoid $W \subset Q$, which has the maximal volume.

Let $G \in \operatorname{int} \mathcal{P}_n$, $v \in \operatorname{int} Q$. We can represent W as follows:

$$\begin{aligned} W &= \{x \in R^n \mid \parallel G^{-1}(x-v) \parallel \leq 1\} \\ &\equiv \{x \in R^n \mid \langle G^{-2}(x-v), x-v\rangle \leq 1\}. \end{aligned}$$

In view of Lemma 4.3.8, the inequality $\langle a, x\rangle \leq b$ is valid for any $x \in W$ if and only if

$$\parallel Ga \parallel^2 \equiv \langle G^2 a, a\rangle \leq (b - \langle a, v\rangle)^2.$$

That gives a convex region for (G, v):

$$\parallel Ga \parallel \leq b - \langle a, v\rangle.$$

Note that $\mathrm{vol}_n W = \mathrm{vol}_n B_2(0,1) \det G$. Therefore our problem can be written as follows:

$$\begin{aligned} &\min_{G,v,\tau} \ \tau, \\ \text{s.t.} \quad & -\ln\det G \leq \tau, \\ & \parallel Ga_i \parallel \leq b_i - \langle a_i, v\rangle, \ i = 1 \dots m, \\ & G \in \mathcal{P}_n, \ v \in R^n, \ \tau \in R^1. \end{aligned} \tag{4.3.15}$$

In view of Lemmas 4.3.7 and 4.3.3, we can use the following self-concordant barrier:

$$\begin{aligned} F(G, v, \tau) &= -\ln\det G - \ln(\tau + \ln\det G) \\ &\quad - \sum_{i=1}^{m} \ln[(b_i - \langle a_i, v\rangle)^2 - \parallel Ga_i \parallel^2], \\ \nu &= 2m + n + 1. \end{aligned}$$

The corresponding efficiency estimate is $O\left(\sqrt{2m+n+1}\cdot \ln \frac{m+n}{\epsilon}\right)$ iterations of a path-following scheme.

4.3.5 Separable optimization

In problems of separable optimization all nonlinear terms are presented by univariate functions. A general formulation of such a problem looks as follows:

$$\begin{aligned} \min_{x\in R^n} \; & q_0(x) = \sum_{j=1}^{m_0} \alpha_{0,j} f_{0,j}(\langle a_{0,j}, x\rangle + b_{0,j}), \\ \text{s.t} \; & q_i(x) = \sum_{j=1}^{m_i} \alpha_{i,j} f_{i,j}(\langle a_{i,j}, x\rangle + b_{i,j}) \leq \beta_i, \; i = 1\dots m, \end{aligned} \tag{4.3.16}$$

where $\alpha_{i,j}$ are some positive coefficients, $a_{i,j}\in R^n$ and $f_{i,j}(t)$ are convex functions of one variable. Let us rewrite this problem in a standard form:

$$\begin{aligned} \min_{x,t,\tau} \; & \tau_0, \\ \text{s.t} \; & f_{i,j}(\langle a_{i,j}, x\rangle + b_{i,j}) \leq t_{i,j}, \; i = 0\dots m, \; j = 1\dots m_i, \\ & \sum_{j=1}^{m_i} \alpha_{i,j} t_{i,j} \leq \tau_i, \; i = 0 \dots m, \\ & \tau_i \leq \beta_i, \; i = 1\dots m, \\ & x\in R^n, \; \tau\in R^{m+1}, \; t\in R^M, \end{aligned} \tag{4.3.17}$$

where $M = \sum_{i=0}^{m} m_i$. Thus, in order to construct a self-concordant barrier for the feasible set of the problem, we need barriers for epigraphs of univariate convex functions $f_{i,j}$. Let us point out such barriers for several important functions.

4.3.5.1 Logarithm and exponent.

Function $F_1(x,t) = -\ln x - \ln(\ln x + t)$ is a 2-self-concordant barrier for the set

$$Q_1 = \{(x,t)\in R^2 \mid x>0, \; t\geq -\ln x\},$$

and function $F_2(x,t) = -\ln t - \ln(\ln t - x)$ is a 2-self-concordant barrier for the set

$$Q_2 = \{(x,t)\in R^2 \mid t\geq e^x\}.$$

4.3.5.2 Entropy function.

Function $F_3(x,t) = -\ln x - \ln(t - x\ln x)$ is a 2-self-concordant barrier for the set

$$Q_3 = \{(x,t) \in R^2 \mid x \geq 0,\ t \geq x\ln x\}.$$

4.3.5.3 Increasing power functions.

Function $F_4(x,t) = -2\ln t - \ln(t^{2/p} - x^2)$ is a 4-self-concordant barrier for the set

$$Q_4 = \{(x,t) \in R^2 \mid t \geq \mid x \mid^p\}, \quad p \geq 1,$$

and function $F_5(x,t) = -\ln x - \ln(t^p - x)$ is a 2-self-concordant barrier for the set

$$Q_5 = \{(x,t) \in R^2 \mid x \geq 0,\ t^p \geq x\}, \quad 0 < p \leq 1.$$

4.3.5.4 Decreasing power functions.

Function $F_6(x,t) = -\ln t - \ln(x - t^{-1/p})$ is a 2-self-concordant barrier for the set

$$Q_6 = \left\{(x,t) \in R^2 \mid x > 0,\ t \geq \tfrac{1}{x^p}\right\}, \quad p \geq 1,$$

and function $F_7(x,t) = -\ln x - \ln(t - x^{-p})$ is a 2-self-concordant barrier for the set

$$Q_7 = \left\{(x,t) \in R^2 \mid x > 0,\ t \geq \tfrac{1}{x^p}\right\}, \quad 0 < p < 1.$$

We omit the proofs of the above statements since they are rather technical. It can be also shown that the barriers for all of these sets, (except maybe Q_4), are *optimal*. Let us prove this statement for the sets Q_6 and Q_7.

LEMMA 4.3.9 *Parameter ν of any self-concordant barrier for the set*

$$Q = \left\{(x^{(1)}, x^{(2)}) \in R^2 \mid x^{(1)} > 0,\ x^{(2)} \geq \tfrac{1}{(x^{(1)})^p}\right\},$$

with $p > 0$, satisfies inequality $\nu \geq 2$.

Proof: Let us fix some $\gamma > 1$ and choose $\bar{x} = (\gamma, \gamma) \in \operatorname{int} Q$. Denote

$$p_1 = e_1, \quad p_2 = e_2, \quad \beta_1 = \beta_2 = \gamma, \quad \alpha_1 = \alpha_2 = \alpha \equiv \gamma - 1.$$

Then $\bar{x} + \xi e_i \in Q$ for any $\xi \geq 0$ and

$$\bar{x} - \beta e_1 = (0, \gamma) \notin Q, \quad \bar{x} - \beta e_2 = (\gamma, 0) \notin Q,$$

$$\bar{x} - \alpha(e_1 + e_2) = (\gamma - \alpha, \gamma - \alpha) = (1,1) \in Q.$$

Therefore, the conditions of Theorem 4.3.1 are satisfied and

$$\nu \geq \tfrac{\alpha_1}{\beta_1} + \tfrac{\alpha_2}{\beta_2} = 2\tfrac{\gamma-1}{\gamma}.$$

This proves the statement since γ can be chosen arbitrarily big. □

Let us conclude our discussion by two examples.

4.3.5.5 Geometric optimization.

The initial formulation of such problems is as follows:

$$\begin{aligned} &\min_{x \in R^n} q_0(x) = \sum_{j=1}^{m_0} \alpha_{0,j} \prod_{j=1}^{n} (x^{(j)})^{\sigma_{0,j}^{(j)}}, \\ &\text{s.t } q_i(x) = \sum_{j=1}^{m_i} \alpha_{i,j} \prod_{j=1}^{n} (x^{(j)})^{\sigma_{i,j}^{(j)}} \leq 1, \; i = 1 \dots m, \qquad (4.3.18) \\ &x^{(j)} > 0, \quad j = 1 \dots n, \end{aligned}$$

where $\alpha_{i,j}$ are some positive coefficients. Note that the problem (4.3.18) is not convex.

Let us introduce the vectors $a_{i,j} = (\sigma_{i,j}^{(1)}, \dots, \sigma_{i,j}^{(n)}) \in R^n$, and change the variables: $x^{(i)} = e^{y^{(i)}}$. Then (4.3.18) is transformed into a *convex* separable problem.

$$\begin{aligned} &\min_{y \in R^n} \sum_{j=1}^{m_0} \alpha_{0,j} \exp(\langle a_{0,j}, y \rangle), \\ &\text{s.t. } \sum_{j=1}^{m_i} \alpha_{i,j} \exp(\langle a_{i,j}, y \rangle) \leq 1, \; i = 1 \dots m. \end{aligned} \qquad (4.3.19)$$

Denote $M = \sum_{i=0}^{m} m_i$. The complexity of solving (4.3.19) by a path-following scheme is

$$O\left(M^{1/2} \cdot \ln \tfrac{M}{\epsilon}\right).$$

4.3.5.6 Approximation in l_p norms.

The simplest problem of that type is as follows:

$$\begin{aligned} &\min_{x \in R^n} \sum_{i=1}^{m} \mid \langle a_i, x \rangle - b^{(i)} \mid^p, \\ &\text{s.t. } \alpha \leq x \leq \beta, \end{aligned} \qquad (4.3.20)$$

where $p \geq 1$. Clearly, we can rewrite this problem in an equivalent standard form:

$$\begin{aligned} & \min_{x,\tau} \tau^{(0)}, \\ \text{s.t } & \mid \langle a_i, x\rangle - b^{(i)} \mid^p \leq \tau^{(i)}, \ i = 1 \dots m, \\ & \sum_{i=1}^{m} \tau^{(i)} \leq \tau^{(0)}, \\ & \alpha \leq x \leq \beta, \\ & x \in R^n, \ \tau \in R^{m+1}. \end{aligned} \tag{4.3.21}$$

The complexity bound of this problem is $O\left(\sqrt{m+n} \cdot \ln \frac{m+n}{\epsilon}\right)$ iterations of a path-following scheme.

We have discussed the performance of interior-point methods on several *pure* problem formulations. However, it is important that we can apply these methods to *mixed* problems. For example, in problems (4.3.7) or (4.3.20) we can treat also the quadratic constraints. To do that, we need to construct a corresponding self-concordant barrier. Such barriers are known for all important examples we meet in practical applications.

4.3.6 Choice of minimization scheme

We have seen that many convex optimization problems can be solved by interior-point methods. However, we know that the same problems can be solved by another general technique, the nonsmooth optimization methods. In general, we cannot say which approach is better, since the answer depends on individual structure of a particular problem. However, the complexity estimates for optimization schemes often help to make a reasonable choice. Let us consider a simple example.

Assume we are going to solve a problem of finding the best approximation in l_p-norms:

$$\begin{aligned} & \min_{x \in R^n} \sum_{i=1}^{m} \mid \langle a_i, x\rangle - b^{(i)} \mid^p, \\ \text{s.t } & \alpha \leq x \leq \beta, \end{aligned} \tag{4.3.22}$$

where $p \geq 1$. And let us have two numerical methods available:

- The ellipsoid method (Section 3.2.6).
- The interior-point path-following scheme.

What scheme should we use? We can derive the answer from the complexity estimates of corresponding methods.

Let us estimate first the performance of the ellipsoid method as applied to problem (4.3.22).

Complexity of ellipsoid method	
Number of iterations:	$O\left(n^2 \ln \frac{1}{\epsilon}\right)$,
Complexity of the oracle:	$O(mn)$ operations,
Complexity of the iteration:	$O(n^2)$ operations.
Total complexity:	$O\left(n^3(m+n)\ln \frac{1}{\epsilon}\right)$ operations.

The analysis of a path-following scheme is more involved. First of all, we should form a barrier model of the problem:

$$
\begin{array}{rl}
 & \min\limits_{x,\tau,\xi} \; \xi, \\
\text{s.t.} & \mid \langle a_i, x\rangle - b^{(i)} \mid^p \le \tau^{(i)}, \; i = 1\dots m, \\
 & \sum\limits_{i=1}^{m} \tau^{(i)} \le \xi, \; \alpha \le x \le \beta, \\
 & x \in R^n, \; \tau \in R^m, \; \xi \in R^1, \\
F(x,\tau,\xi)) = & \sum\limits_{i=1}^{m} f(\tau^{(i)}, \langle a_i, x\rangle - b^{(i)}) - \ln(\xi - \sum\limits_{i=1}^{m} \tau^{(i)}) \\
 & - \sum\limits_{i=1}^{n} [\ln(x^{(i)} - \alpha^{(i)}) + \ln(\beta^{(i)} - x^{(i)})],
\end{array}
\qquad (4.3.23)
$$

where $f(y,t) = -2\ln t - \ln(t^{2/p} - y^2)$.

We have seen that the parameter of barrier $F(x,\tau,\xi)$ is $\nu = 4m+n+1$. Therefore, the number of iterations of a path-following scheme can be estimated as $O\left(\sqrt{4m+n+1}\cdot \ln \frac{m+n}{\epsilon}\right)$.

At each iteration of the path-following scheme we need to compute the gradient and the Hessian of barrier $F(x,\tau,\xi)$. Denote

$$
g_1(y,t) = f'_y(y,t), \quad g_2(y,t) = f'_t(y,t).
$$

Then

$$F'_x(x,\tau,\xi) = \sum_{i=1}^m g_1(\tau^{(i)}, \langle a_i, x\rangle - b^{(i)})a_i - \sum_{i=1}^n \left[\tfrac{1}{x^{(i)}-\alpha^{(i)}} - \tfrac{1}{\beta^{(i)}-x^{(i)}}\right] e_i,$$

$$F'_{\tau^{(i)}}(x,\tau,\xi) = g_2(\tau^{(i)}, \langle a_i, x\rangle - b^{(i)}) + \left[\xi - \sum_{j=1}^m \tau^{(j)}\right]^{-1},$$

$$F'_\xi(x,\tau,\xi) = -\left[\xi - \sum_{i=1}^m \tau^{(i)}\right]^{-1}.$$

Further, denoting

$$h_{11}(y,t) = f''_{yy}(y,t), \quad h_{12}(y,t) = f''_{yt}(y,t), \quad h_{22}(y,t) = f''_{tt}(y,t),$$

we obtain

$$\begin{aligned}
F''_{xx}(x,\tau,\xi) &= \sum_{i=1}^m h_{11}(\tau^{(i)}, \langle a_i, x\rangle - b^{(i)})a_i a_i^T \\
&\quad + \text{diag}\left[\tfrac{1}{(x^{(i)}-\alpha^{(i)})^2} + \tfrac{1}{(\beta^{(i)}-x^{(i)})^2}\right], \\
F''_{\tau^{(i)}x}(x,\tau,\xi) &= h_{12}(\tau^{(i)}, \langle a_i, x\rangle - b^{(i)})a_i, \\
F''_{\tau^{(i)},\tau^{(i)}}(x,\tau,\xi) &= h_{22}(\tau^{(i)}, \langle a_i, x\rangle - b^{(i)}) + \left(\xi - \sum_{i=1}^m \tau^{(i)}\right)^{-2}, \\
F''_{\tau^{(i)},\tau^{(j)}}(x,\tau,\xi) &= \left(\xi - \sum_{i=1}^m \tau^{(i)}\right)^{-2}, \; i \neq j,
\end{aligned}$$

$$F''_{x,\xi}(x,\tau,\xi) = 0, \quad F''_{\tau^{(i)},\xi}(x,\tau,\xi) = -\left(\xi - \sum_{i=1}^m \tau^{(i)}\right)^{-2},$$

$$F''_{\xi,\xi}(x,\tau,\xi) = \left(\xi - \sum_{i=1}^m \tau^{(i)}\right)^{-2}.$$

Thus, the complexity of the second-order oracle in the path-following scheme is $O(mn^2)$ arithmetic operations.

Let us estimate now the complexity of each iteration. The main source of computations at each iteration is the solution of a Newton system. Denote

$$\kappa = \left(\xi - \sum_{i=1}^m \tau^{(i)}\right)^{-2}, \quad s_i = \langle a_i, x\rangle - b^{(i)}, \; i = 1 \ldots n,$$

and

$$\Lambda_0 = \mathrm{diag}\left[\frac{1}{(x^{(i)}-\alpha^{(i)})^2} + \frac{1}{(\beta^{(i)}-x^{(i)})^2}\right]_{i=1}^n \quad \Lambda_1 = \mathrm{diag}\,(h_{11}(\tau^{(i)}, s_i))_{i=1}^m,$$

$$\Lambda_2 = \mathrm{diag}\,(h_{12}(\tau^{(i)}, s_i))_{i=1}^m, \quad D = \mathrm{diag}\,(h_{22}(\tau^{(i)}, s_i))_{i=1}^m.$$

Then, using the notation $A = (a_1, \ldots, a_m)$, $e = (1, \ldots, 1) \in R^m$, the Newton system can be written in the following form:

$$\begin{aligned} &[A(\Lambda_0 + \Lambda_1)A^T]\Delta x + A\Lambda_2\Delta\tau = F'_x(x, \tau, \xi), \\ &\Lambda_2 A^T\Delta x + [D + \kappa I_m]\Delta\tau + \kappa e\Delta\xi = F'_\tau(x, \tau, \xi), \\ &\kappa\langle e, \Delta\tau\rangle + \kappa\Delta\xi = F'_\xi(x, \tau, \xi) + t, \end{aligned} \tag{4.3.24}$$

where t is the penalty parameter. From the second equation in (4.3.24) we obtain

$$\Delta\tau = [D + \kappa I_m]^{-1}(F'_\tau(x, \tau, \xi) - \Lambda_2 A^T\Delta x - \kappa e\Delta\xi).$$

Substituting $\Delta\tau$ in the first equation in (4.3.24), we can express

$$\begin{aligned} \Delta x &= [A(\Lambda_0 + \Lambda_1 - \Lambda_2^2[D + \kappa I_m]^{-1})A^T]^{-1}\{F'_x(x, \tau, \xi) \\ &\quad -A\Lambda_2[D + \kappa I_m]^{-1}(F'_\tau(x, \tau, \xi) - \kappa e\Delta\xi)\}. \end{aligned}$$

Using these relations we can find $\Delta\xi$ from the last equation in (4.3.24).

Thus, the Newton system (4.3.24) can be solved in $O(n^3 + mn^2)$ operations. This implies that the total complexity of the path-following scheme can be estimated as

$$O\left(n^2(m+n)^{3/2} \cdot \ln \frac{m+n}{\epsilon}\right)$$

arithmetic operations. Comparing this estimate with that of the ellipsoid method, we conclude that the interior-point methods are more efficient if m is not too large, namely, if $m \leq O(n^2)$.

Bibliography

Chapter 1. Nonlinear optimization

Section 1.1. Complexity theory for black-box optimization schemes was developed in [8]. In this monograph the reader can find different examples of resisting oracles and lower complexity bounds similar to that of Theorem 1.1.2.

Sections 1.2 and 1.3. There exist several classic monographs [2, 3, 7], which treat different aspects of nonlinear optimization and numerical schemes. For sequential unconstrained minimization the best source is still [4].

Chapter 2. Smooth convex optimization

Section 2.1. The lower complexity bounds for smooth convex and strongly convex functions can be found in [8]. However, the proof used in this section seems to be new.

Section 2.2. Gradient mapping was introduced in [8]. The optimal method for smooth and strongly smooth convex functions was proposed in [10]. A constrained variant of this scheme is taken from [11].

Section 2.3. Optimal methods for minimax problems were developed in [11]. The approach of Section 2.3.5 seems to be new.

Chapter 3. Nonsmooth convex optimization

Section 3.1. A comprehensive treatment of different topics of convex analysis can be found in [5]. However, the classic [15] is still very useful.

Section 3.2. Lower complexity bounds for nonsmooth minimization problems can be found in [8]. The scheme of the proof of the convergence rate was suggested in [9]. See [13] for detailed bibliographical comments on the history of nonsmooth minimization schemes.

Section 3.3. The example for the Kelley method is taken from [8]. The presentation of the level method is close to [6].

Chapter 4. Structural optimization

This chapter contains an adaptation of the main concepts from [12]. We added several useful inequalities and slightly simplified the path-following scheme. We refer the reader to [1] for numerous applications of interior-point methods, and to [14], [16], [18] and [19] for detailed treatment of different theoretical aspects.

References

[1] A. Ben-Tal and A. Nemirovskii. *Lectures on Moden Convex Optimizatin Analysis, Alogorithms, and Engineering Applications*, SIAM, Philadelphia, 2001.

[2] A.B. Conn, N.I.M. Gould and Ph.L. Toint. *Trust Region Methods*, SIAM, Philadelphia, 2000.

[3] J.E. Dennis and R.B. Schnabel. *Numerical Methods for Unconstrained Optimization and Nonlinear Equations*, SIAM, Philadelphia, 1996.

[4] A.V. Fiacco and G.P. McCormick. *Nonlinear Programming: Sequential Unconstrained Minimization Techniques*, John Wiley and Sons, New York, 1968.

[5] J.-B. Hiriart-Urruty and C. Lemaréchal. *Convex Analysis and Minimization Algorithms*, vols. I and II. Springer-Verlag, 1993.

[6] C. Lemarechal, A. Nemirovskii and Yu. Nesterov. New variants of bundle methods. *Mathematical Programmming*, 69(1995), 111–148.

[7] D.G. Luenberger. *Linear and Nonlinear Programming*, Second Edition, Addison Wesley. 1984.

[8] A.Nemirovsky and D.Yudin. *Informational complexity and efficient methods for solution of convex extremal problems*, Wiley, New York, 1983.

[9] Yu.Nesterov. Minimization methods for nonsmooth convex and quasiconvex functions. *Ekonomika i Mat. Metody*, v.11, No.3, 519-531, 1984. (In Russian; translated as *MatEcon.*)

[10] Yu.Nesterov. A method for solving a convex programming problem with rate of convergence $O(\frac{1}{k^2})$. *Soviet Math. Doklady*, 1983, v.269, No.3, 543-547. (In Russian.)

[11] Yu.Nesterov. *Efficient methods in nonlinear programming.* Radio i Sviaz, Moscow, 1989. (In Russian.)

[12] Yu. Nesterov and A.Nemirovskii. *Interior-Point Polynomial Algorithms in Convex Programming*, SIAM, Philadelphia, 1994.

[13] B.T. Polyak. *Introduction to Optimization.* Optimization Software, New York, NY, 1987.

[14] J.Renegar. *A Mathematical View of Interior-Point Methods in Convex Optimization*, MPS-SIAM Series on Optimization, SIAM 2001.

[15] R.T. Rockafellar *Convex Analysis*, Princeton Univ. Press, Princeton, NJ, 1970.

[16] C. Roos, T. Terlaky and J.-Ph. Vial. *Theory and Algorithms for Linear Optimization: An Interior Point Approach.* John Wiley, Chichester, 1997.

[17] R.J. Vanderbei. *Linear Programming: Foundations and Extensions.* Kluwer Academic Publishers, Boston, 1996.

[18] S. Wright. Primal-dual interior point methods. SIAM, Philadelphia, 1996.

[19] Y. Ye. *Interior Point Algorithms: Theory and Analysis*, John Wiley and Sons, Inc., 1997.

Index

www.ingramcontent.com/pod-product-compliance
Lightning Source LLC
Chambersburg PA
CBHW050808100925
32286CB00003BA/207
* 9 7 8 1 4 0 2 0 7 5 5 3 7 *